NATTERING ON THE NET

Women, Power and Cyberspace

Dale Spender

SPINIFEX

Spinifex Press Pty Ltd
504 Queensberry Street
North Melbourne, Vic. 3051
Australia

First published by Spinifex Press, 1995

Typeset in 11/14 pt Times by
    Claire Warren, Melbourne
Indexed by Max McMaster
Made and printed in Australia by
    Australian Print Group
Cover design by Liz Nicholson, Design Bite

National Library of Australia
Cataloguing-in-Publication data:

Spender, Dale.
    Nattering on the net: women, power and
    cyberspace.

    Bibliography.
    Includes index.
    ISBN 1 875559 09 4.

    1. Computer networks – Social aspects.
    2. Internet (Computer network) – Social
    aspects. 3. Women – Social conditions.
    I. Title

303.4834082

For Cheris Kramarae

Contents

Acknowledgements

This book would never have been started without the support – and the stirring – of Cheris Kramarae. At a time when I dismissed computers as no more than "glamour cut-and-paste", she enlisted the help of Eudora (an e-mail system) and persuaded me of its possibilities.

For almost twenty years we have worked together on a variety of projects. We were the founding editors of *Women's Studies International Quarterly*; the co-editors of *The Knowledge Explosion*; with H. Jeanie Taylor, the founding members of Women, Information Technology and Scholarship (WITS); and the co-originators of *Women's International Knowledge: Education and Data* (WIKED), an international database. We are now in the process of putting together a treasure-chest of resources for feminist CD-ROMs.

Cheris Kramarae has been a central part of my professional and personal life, and for the good friend she has been to me, and to the international women's movement, I thank her, and pay tribute to her wisdom and her unfailing generosity.

What Cheris Kramarae started in relation to computers, H. Jeanie Taylor nurtured and organised. As Associate Director of the Center for Advanced Studies at the University of Illinois, she has been the energising force behind WITS. I am more than grateful for the seminars she has set up, the projects she has initiated, and the counsel and care she has given me over countless cups of cappuccino royale in Urbana Champaign.

Visionary aims and enthusiasm, however, are not enough to turn an old hand at lit. crit. into a data crit. advocate. For guiding me – and my Mac – through cyberspace, I am indebted to Rick Ernst. He is one of those rare young men who not only knows the techie stuff, but is virtually magic in the way he empowers others to become computer proficient.

Then there is my sister, Lynne Spender, who says she has read every word. More than once. She provides daily affirmation and advice. And her help is not confined to feedback on my prose style. Her sense of humour, and her perspective on the world and on writing, have entertained me and informed me. Her expertise in the area of authorship and writers' rights in an electronic society has been invaluable in my research and my writing life.

Parts of this manuscript have also been read carefully by Ted Brown, and commented upon accordingly. Much of the cost of my conversion to computers has been borne by him. He does a great deal to keep my world in order, and I am enormously grateful to him for all his contributions.

My neighbour, Quentin Bryce, has been an inspiration in more ways than one. Apart from the snacks and the solace that she unstintingly provides, her complaint that it was getting embarrassing when people asked her when this book would be finished certainly spurred me on.

Sue Hawthorne and Renate Klein have also been a tower of strength throughout the process of research, writing and publishing. I thank them for their patience; and I would also like to thank Kelly McElroy, who, in the absence of an intelligent agent, has forwarded to me a steady stream of press clippings on computers and the new technologies.

Without the help of Sarah Neal, this work would have taken much longer. Not only has she held the fort on numerous occasions, but her research skills and her information presentation are out of this world. As a successful young film-maker, she too has helped me make the transition to the new media, and has expanded my horizons.

Libby LeClair has made my life more manageable and civilised. For all the time and effort that has gone into writing briefs and drawing up schedules, I am deeply appreciative. It's true that I couldn't manage without her.

Nor would I want to continue without the savvy and sophistication of Jill Hickson and Associates. Agents are rarely acknowledged for the perceptive analysis and constructive criticism they provide, but Jill Hickson does even more than this; I am ever grateful for her commitment and her competence.

I would also like to thank Gail Furness for her friendship, food and wine flair, and her thought-provoking professional insights.

My mother and father, Ivy and Harry Spender, have made a significant contribution to this work. They too have regularly clipped appropriate articles from newspapers and have forwarded them to me all over the world. This is not the least of their support services, but it is the most recent.

Kirsten Lees has been a colleague and a valued friend. Her experience of the publishing industry and her analysis of electronic trends have been extremely useful. I hope that there will be future projects on which we can work together.

I would also like to thank Helen Thomson for her facility with faxes, and for sharing her experiences of writing for deadlines. We have enjoyed many good meals and solved many problems over the years. I have benefited greatly from her friendship and her literary and journalistic expertise.

Eleanor Ramsay has been everything you could want in a best friend. For her unqualified praise, complete faith and support, her astute analysis and unswerving loyalty, I owe her a great deal. I am more than appreciative of all her efforts.

Roving researcher Robyn Daniels has made a major contribution to my work. She has sent articles, left messages about relevant programs, and constantly kept in touch. I value her understanding, her energy, and her principles.

In the computer world I have relied greatly on the expertise and support of Rosie Cross and Virginia Barratt in Australia. They are "geekgirls" whose computer competence speaks for itself and serves as a model for the next generation. I also want to pay tribute to the many skills and the great goodwill of Patricia Gillard, Jen St Clair, Julie James Bailey, and Elizabeth Reid.

In the United States I am grateful to Maureen Ebben, Mary Hocks, Gail Hawisher, Kal Alston, Karen Helleyer, Frances Jacobsen, and all the WITS women, who, along with Amy Bruckman from MIT, have shared information and provided suggestions and encouragement.

I am also indebted to the University of Queensland, and not just for computer services. Ian O'Connor and Norman Smith have been a constant source of inspiration and assistance; and Linda Rosenman and Art Shulman have regularly provided a friendly context for the exchange of ideas. Janine Schmidt, Jan Reid, and Ian Reineke have been the very best of professional colleagues.

For their excellent and extensive contribution to the working party on electronic scholarship, I would like to thank Kim Hosking and Gilgun Cribb; the issues that are raised could never have been so systematically addressed without their assistance.

Margie Byrne (of the NSW State Library) and Alison Crook have gone out of their way to keep me up to date with the latest technological developments: and David Loader, of Methodist Ladies' College, Melbourne, has kindly allowed me to observe classes at the school, and to comment on the transition that is taking place in education.

The Hon. Barry Jones has not only helped to pave the way for the discussion of the new technologies, but has set out stimulating and critical comment on directions for research. Paige Porter has always given me the benefit of her clear, perceptive insights about education and life; and McKenzie Wark has not only made his own valuable contribution to the area, but has been for me an excellent sounding-board .

Robert Pullan and Ross Fitzgerald continue to try to persuade me of the benefits of their version of free speech, and I have thoroughly enjoyed our discussions on censorship and the Internet. I am also grateful to Anne Deveson for her vision and her management experience.

The friendship and the skills of Lance Mason have done much to improve my quality of life, as has the support of my friends Janet Irwin, Cathryn Mittelheuser and Margaret Mittelheuser. I am grateful to Katy Steenstrup and Jan Owen for the invigorating contribution they make, and to Rosie Scott, Pam Gilbert, Sara Hardy, and Mary Crawford for their psychological sustenance.

And I owe much to Colleen Forrester for her friendship, her example, and her astute analysis of the way the world works.

I would also like to pay tribute to the work of Lisa Bellear, who has made an immeasurable difference to my own education. She has shown great spirit and generosity in building bridges across cultural and material divides; it is her commitment to the principles of access and equity that stand as a symbol for all of us in the new information world.

Dale Spender
July 1995

Introduction

This is not a book about computers. It is a book about people. It's about the impact that computers are having on human society. The reason for this focus is that who we are, what we know, and how we think, are all being changed as we move from a print-based society to a computer-based world. We are becoming different people; we are creating a new community.

So far, there has been little discussion about this cyber-community and how it affects us. Most of the talk has been in relation to the technology; the marvels of the "chip", the power of computers, and the vast potential of cyberspace. But, given the consequences of technology in the past, we can't just assume that all will be well.

In describing the possibilities of cyberspace, use is often made of a highway metaphor. Just as highways opened up the country to everyone in Australia and the United States, for example, so too, by implication, the information superhighway will open up the electronic world of communication, education, entertainment and services. But, as William Howell points out in *The Chronicle of Education*, the promise of access, mobility, and a better way of life has not always been delivered, and we should be wary of the hype. In the USA, he says, citizens paid a lot for the so-called benefits:

> by promoting the interests of petroleum producers and automobile manufacturers ahead of public transportation, we saddled ourselves with side effects such as traffic deaths, air pollution and urban gridlock.[1]

Not to mention urban sprawl, isolated suburbs, and the downtown slums. And, he adds, with a sense of foreboding:

> Without some major changes now in our approach to the information revolution, we'll pay heavily again and on a much grander scale.[2]

---

1. William Howell, 1994, "Point of View", *Chronicle of Higher Education*, 8 June, p. A40.
2. Howell, "Point of View", p. A40.

It's easy to be carried away by all the good news about the superhighway and what it will mean for the wealthier and developed societies, but a little bit of doubt won't go astray. Already some of the pioneers are suggesting that there should be increased reflection and responsibility.

Clifford Stoll, a computer geek from the earliest days (the man who caught the German spies hacking through the system), is having second thoughts, for example. Not that he is against the technology – he just wants to see a little more caution exercised when it comes to putting the information infrastructure in place. In *Silicon Snake Oil* he introduces his argument with the statement: "Computers themselves don't bother me, it's the culture in which they're enshrined."[3]

Computers don't bother me either. In fact, I have to acknowledge that I am a complete convert. My computer means as much to me now as a library of books once did – it represents a key to another universe, to a realm of information, creativity, and international ideas. But, like Clifford Stoll, I am concerned about the culture and the effects that the electronic revolution is having on society. I want to see computers, culture and communities as a major topic on the public agenda.

I am sure that all the discussion on broadband versus narrowcasting, rams, bytes and fibre optics deals with important issues, and is of immense fascination to some members of the population. But these are no more the substance of the electronic revolution than the emergence of the factory was the industrial revolution; in both cases it is the change in society – the shifts in power, wealth, influence, organisation, and the environmental consequences – that matters to us all as individuals, and as communities.

Our priority now should be to put human beings at the centre of computer culture; we must start thinking, planning, and managing the information revolution. Because every social issue that we are familiar with in the real world will now have its counterpart in the virtual one.

Everything from sexual harassment to questions of the distribution of wealth and power, from concepts of privacy to provision for access and equity, will all be up for policy discussion and decision-making. As Professor Ian O'Connor has perceptively stated, "What we now need is a Department of Social Policy for Cyberspace."[4]

The emergence of cyberspace challenges the horizons and the habits of print-based culture. It is now more than five hundred years since the printing press was introduced, and with it came a social revolution in the Western world and the foundations of contemporary society. In its own way, print helped to construct the particular Western notions of individual and community (a conceptualisation not always shared by the Chinese, for example). So engrained is this print-world view that we aren't necessarily conscious of the hold it has had on our minds. As Neil Postman says,

---

3. Clifford Stoll, 1995, *Silicon Snake Oil: Second Thoughts on the Information Superhighway*, Bantam Doubleday, New York, p. 3.
4. Ian O'Connor, Professor of Social Work and Social Policy, University of Queensland, private communication.

A person who reads a book or who watches television or who glances at his [*sic*] watch is not usually interested in how his mind is organized or controlled by these events, still less in what idea of the world is suggested by a book, television, or a watch.[5]

Just as a watch and a book have influenced who we are and how we explain our world, so now are the new technologies "reprogramming" the human condition.

For we are probably the last of the purely print-proficient.

We are the last generation to be reared within a culture in which print is the primary information medium. Because we have grown up and become skilled in a print-based community, we have developed certain ways of making sense of the world. We are, to some extent, what print has made us. And now we have to change.

Not surprisingly, we are a generation prone to suffer from acute information anxiety.

There have been earlier information revolutions, of course. The shift from an oral culture to a written culture was just as decisive and dramatic as is the shift we are experiencing today. But it took much longer for the storytellers of oral times to make way for the scribes; it was a change that could be measured in terms of centuries. The current shift is taking place within the space of one lifetime. And we are not a society which has been well prepared for such a pressured pace of change.

Certain values have served us well while we have used print to make our world. They are not as useful – and can actually be an obstacle – now that we are obliged to work with the new technologies. This increasingly places many of us in the uncomfortable position of being deskilled.

All the expertise that we have acquired with print no longer commands the respect that it did when we started our working lives. So much of our living and working experience is already computer-based that many of us are on a constant steep learning curve as we keep adjusting to technological innovations.

"I have just learnt to do e-mail," one of my friends who runs a bookshop said with pride – and tension. "Can I stop now?" she asked, "Am I there yet? Can this be it?" While I know she speaks for many, the answer has to be no; there is no end in sight. For those of us who were reared with print, the continual effort to learn the new technologies will be an ongoing fact of life.

Whether it is the academic men in the university who have had to learn to do their own typing, or the older person who has had to come to grips with money machines, voice mail or microwave ovens, the shift to the computer has to be made. Not only will the trend persist: the pace is likely to increase. As Nicholas Negroponte says: "Computing is not about computers anymore. It is about living." He goes on:

computers are moving into our daily lives: 35% of American families and 50% of American teenagers have a personal computer at home; 30 million people are estimated to be on the Internet; 65% of new computers sold worldwide in 1994 were for the home; and 90% of those to be sold this year are expected to have modems and CD-ROM drives. These numbers

5. Neil Postman, 1986, *Amusing Ourselves to Death: Public Discourse in the Age of Show Business*, Penguin, New York, p. 11.

do not even include the fifty microprocessors in the average 1995 automobile, or the micro-processors in your toaster, thermostat, answering machine, CD player and greeting cards. And if I am wrong about any of the numbers above, just wait a minute.[6]

There are problems and difficulties in all the personal, intellectual and work changes that we are being required to make, but this is only one side of the story. There are opportunities as well. What we have to see is that we are the only generation which will know both mediums, the print and the electronic. We are the ones who will be able to make comparisons, who will be able to assess, evaluate and transfer our experience, expertise – and wisdom – from the old forms to the new.

We have a great deal to offer. The only condition is that we become computer-proficient. We must get to know our way around the computer world in the same way that we have been at home with print. At one level – assuming the availability of resources – this goal isn't difficult to achieve.

Computer-competency is not an option any more. It is a condition of citizenship in the electronic world. This is why particular emphasis is paid to women and computers in this book. For many reasons – which have less to do with women and more to do with computers – women are not making the shift to the new medium at the same rate as men: the most recent reliable figures indicate that 94 per cent of Internet users are male.[7]

Despite the belief of some individuals, the computer is not a toy; it is the site of wealth, power and influence, now and in the future. Women – and Indigenous people, and those with few resources – cannot afford to be marginalised or excluded from this new medium. To do so will be to risk becoming the information-poor. It will be to not count; to be locked out of full participation in society in the same way that illiterate people have been disenfranchised in a print world.

There are people who say they cannot cope with all this change. Some think the computer is alien and evil ("I'd rather be dead than use one of those things," an estab-lished author snarled at me), and some firmly believe that the skills they have will see them out. But, unfortunate and unjust as it is, they are making a monumental mistake if they hang on to such attitudes. Anyone planning to be around in the next few years cannot cling to the ideas that print skills will be enough, and that being computer-incompetent will be no great loss.

None of us can choose to stay outside the computer culture. (Some of us will be forced outside, and that is a very different matter.) Indeed, it will be much harder and more distressing to stay out than it will be to get in to the new ways of doing and being.

Becoming computer-competent is not all that difficult. Learning to read (in print culture) is probably more complex, and learning to drive a car is undeniably more dan-gerous. As we know full well, three-year-olds can manage the new technologies, often with much more facility than can their parents. This is not because the three-year-olds

---

6. Nicholas Negroponte, 1995, *Being Digital*, Knopf, New York, pp. 5–6.
7. James Pitkow and Mimi Recker, Georgia Institute of Technology: see Donald Carli, 1994, "A Designer's Guide to the Internet", *Step-by-Step Graphics*, Peoria, Ill., November–December, p. 27.

are brighter or more technologically gifted. Computers are not technology to them; they are just the way their world works. Computers just happened to be there when they were born; for the rest of us, *technology is what wasn't invented when we came into the world.*

For those of us who have to adjust to the new technology, the message should be loud and clear. We can start with the recognition that a computer is, to some extent, nothing other than an enhanced telephone, and it need be no more difficult to use. It's just that some of us need a lot of counselling before we can appreciate this. With a little bit of help from our friends (most women who are computer-competent were introduced to the new technology by a friend), we can get over our reticence and take to nattering on the net with the same comfort and ease – and sense of satisfaction – as we took up the telephone not so long ago. For many of us who are print-proficient, it is not technological information we need to make the move – it is confidence.

I don't see the tap in the bathroom – or even the book on the shelf – as technology, although they could both be mystifying objects to any adult who had never seen them before. The technological elements of the computer will also be increasingly invisible in the future to those who grow up with computers and who are computer-competent.

This is not just because computers will be so much part of the atmosphere; it is also because the technology itself will be more sophisticated and more subtle. There is no better example of this infiltration of our daily lives than that of the intelligent agent. This is where we are heading, and it is an indication of how we are going to change our thinking and our activities.

Almost everyone who has logged into cyberspace has become aware of the vast amount of undifferentiated information that is available now that so many millions of computers are connected, and so many people can put out the information, or "publish", whatever they please. As there are no teachers, librarians, or sages to help you make your way around on the information superhighway, the most common complaint is that it is too hard to find what you need.

This is why we have to take computers one step further. In the words of Nicholas Negroponte, the founding director of the famous MultiMedia Lab at MIT, we have to create computers which will "filter, sort, prioritize and manage multimedia on our behalf – computers that read newspapers and look at television for us, and act as editors when we ask them to do so".[8] These new computers are called *intelligent agents.* Not surprisingly, "interest in intelligent agents has become the most fashionable topic of research in human interface design".[9] Here is the signpost to cyberspace.

These intelligent agents can live in two places. They can live – and sift and sort – at the point of origin, or they can live and work at the point of reception. I am not all that fussed about whether the agent lives at someone else's place, or mine. Either way, I can hardly wait to have one working for me.

---

8. Negroponte, *Being Digital,* p. 20.
9. Negroponte, *Being Digital,* p. 151.

Take newspapers, for example. An agent will let me read exactly what I want. It will mean an end to all those football grunts and jerks: an end to any violent sport (such as boxing); indeed, an end to any game that I might find offensive. I could even request my intelligent agent to transmit to me only women's sport, or any good news of the day.

If the agent resides at the *Courier Mail* office, this will be like having my own staff reporters. I will get on my screen each morning (and I can download if I wish), a newspaper that has been custom-made for me. This is one of the ironies of the so-called new *mass* media: the audience may often be just *one* person!

If, however, the agent lives at my place, it could be that during the night a number of different newspapers send out their information, and my agent samples all of them, looking for anything I might like. Then it provides me with a personalised coverage from this wide range of sources. Oh what joy to be able to manage my own news. And what implications it will have for newspapers. Not only could women's newspapers be constructed for the first time, but newspaper editors might start to see the potential of such a niche market, and set about making news which is of greater concern to women. Good news for all.

Roll on the next stage of computer interfaces.

There is no reason that I cannot have an agent working for me at such places as the *Courier Mail*, the movie transmission centre, and Radio National. Each source could be routinely filtered for my preferences. Not only will I be able to see just what I choose: I will be able to do it at whatever time suits me:

> The six o'clock news not only can be delivered when you want it, but it can also be edited for you and randomly accessed by you. If you want an old Humphrey Bogart movie at 8.17 p.m., the telephone company can provide it . . . Eventually when you watch a baseball game, you will be able to do so from any seat in the stadium, or for that matter from the perspective of the baseball.[10]

Of course these are Mr Negroponte's choices; my own agent would be given very different instructions to follow. As would everyone else's.

Intelligent agents will be searching global information and presenting it in a way that makes sense to each person who has the resources to be connected (let's not lose sight of this throughout)! Some people have likened an agent to a press-clipping service, where the particular brief is to clip items of interest to each individual from the masses of data that are floating around. Others have suggested that an intelligent agent is more like an English butler.

To Nicholas Negroponte, the expert on matching computers with human needs, it is the well-trained English butler that provides the model for the next stage in interfaces. An agent will do a great deal more than set you up with a newspaper service and sort your e-mail messages.

---

10. Negroponte, *Being Digital*, p. 49.

The "agent" answers the phone, recognizes the callers, disturbs you when appropriate, and may even tell a white lie on your behalf. The same agent is well trained in timing, versed in finding the opportune moments, and respectful of idiosyncrasies.[11]

To the monks of the Middle Ages who read but a few manuscripts in their lifetime, the avid book reader probably looked like a superhuman creature; that one person could have access to so much information from so many books was beyond their comprehension. The cyber-person looks much the same to those of us who have been used to the slower and more cumbersome process of print; how could one person cope with so much data – and what could they do with it? What sort of life will you have in this high-tech cyberworld, being served by so many British-butler intelligent agents?

One significant change in the daily scenery that the intelligent agent will bring is the shift away from the television to the computer screen. When it comes to multimedia, a screen is a screen is a screen . . . As Nicholas Negroponte points out, the consensus in the industry is that the distinction between the two types of screen will soon disappear; the one that is the gateway to cable, telephone or satellite will be the one that we use.

This will be more like what we think of now as the PC than the television. "In other words," says Nicholas Negroponte, "there is no TV-set industry in the future. It is nothing more or less than a computer industry."[12] We won't be buying TV sets from the shops. If we do continue to call one of the screens a television, it will be because of where we place it in the house. And perhaps how far away we sit from it.

The point is that televisions are not very "intelligent" machines. You can't *interact* with televisions as we know them at the moment. While you can zap around the channels (and record with a VCR), you are still the one who has to do the choosing, and you still have to take what you can get. Even with pay or cable TV. For those reared within computer culture, this form of passivity (or absence of an intelligent interface) will not be good enough. They are not going to wait for the weather report for example; with the help of their agent and their computer-television screen, they are going to "do" their own information.

Instead of broadcasting the weatherman and his proverbial maps and charts, think of sending a computer model of the weather. These bits arrive in your computer-TV and then you, at the receiving end, implicitly or explicitly use local computing intelligence to transform them into a voice report, a printed map, or an animated cartoon with your favorite Disney character. The smart TV set will do all this in whatever way you want, maybe even depending on your disposition and mood at the moment. In this example the broadcaster does not even know what the bits will turn into: video, audio or print. You decide that.[13]

To those of us who have been educated and entertained with books, the medium of print may seem far from passive. But to the products of the new computer culture,

11. Negroponte, *Being Digital*, p. 150.
12. Negroponte, *Being Digital*, p. 47.
13. Negroponte, *Being Digital*, p. 55.

print will appear as fixed, unwieldy, nonactive; even dull and boring. (Such charges are already being levelled at print in our education system.) The next generation will not be content to sit and read or watch someone else's information all the time; they will want to construct their own version.

That we are moving from being "readers" to "users" is one of the themes that flows through this book. The starting point for the discussion is print culture itself, with all its characteristics of stability, order and standardisation.

Even a quick survey suggests that many aspects of print culture that many people have held to be sacred are nothing other than conventions, and open to change. Supposedly proper grammar, dictionary definitions and correct spelling have all been a result of printing press uniformity. The standardised forms did not exist before the advent of the printing press, and they probably won't exert the same hold over our minds as the influence of the printing press declines.

It is interesting to note just how upset some people are by the very idea of a return to multiple spellings, for example. Such responses say more about our fear of change than about the role of spelling in creating the social fabric, or the meaning of life: I am sure we will all readily survive spelling diversity.

Chapter 1 not only outlines the set characteristics of print culture which we have come to take for granted, but points to the similarities between the print information revolution of the fifteenth century and the present information revolution. The medieval monks who created and valued those precious manuscripts were very much against the introduction of books. Many of the objections they raised then to the spread of print are the very same ones being used again now to discredit electronic communication.

The protests against print included everything from accusations that books were mindless, that they appealed to the lowest common denominator, that they were only produced to make a profit, right through to arguments that the book would ruin conversation in the family, and would lower the standards of civilisation.

These historical details are useful to us. Once we are aware of the parallels between then and now, we are in a better position to determine how far our current reservations are related to the medium itself, and how far they are the negative reactions of a community where the culture is undergoing rapid change.

Chapter 2 takes up the issue of literature. Is it the end of the novel? The end of the great masters? Will we lose the words of the great white men who have formed our literary tradition since the printing press was invented? Who will win the struggle for the word – as portrayed by David Williamson in his excellent play, *Dead White Males*?

Some high-culture parts of the print tradition won't carry over to the twenty-first century. (Some have already been thrown out, if you believe the more conservative academics, like Allan Bloom, for example.)[14] For the people who have held these writers to be centrally important to their own development and that of society, this will be a great loss.

---

14. Allan Bloom, 1987, *The Closing of the American Mind*, Penguin, New York.

There is no denying this. It will be as it was for the monks when the rise of the book meant the loss of those beautiful, illuminated manuscripts that were the repository of human wisdom. In retrospect, few of us would believe that society lost out, that humanity was worse off once the book made its appearance. It's a matter of weighing the gains against the losses.

No doubt, future generations will do much the same thing; they will think that the disappearance of some texts is not such a high price to pay for the wealth of the information revolution. Besides, the literary canon has always been based on selection, with the works of many writers systematically being lost from one generation to the next. This is why women have disappeared, and why it has been such a white male canon. Some of the men will now go the same way as the many brilliant women and black literary figures have gone in the past; this could be regarded as no great tragedy in some intellectual circles.

Chapter 3 is about the habit of reading; about when it started, how it is done, and where it is going. It can be a shock to learn that only five hundred years ago, reading was held to be such a special skill that only the very talented few could ever hope to be able to do it. Reading then was a very different activity from what it has been for the last hundred years, and from what it is now.

Print led to the democratisation of reading; it enabled the many to read, rather than just the few. The masses were empowered in a way that had not been possible during the manuscript period. In fact, so important has print been for the last five hundred years, that literacy has come to be regarded as a human right.

Books have been highly valued; they have come to represent more than the sum of their parts. Countless numbers of readers have seen them as the gateway to ideas, information and imagination – to another realm of existence. In the eighteenth and nineteenth centuries, when women were barred from educational institutions, books were the lifeline for many. Florence Nightingale was just one who declared that without the book, women would be in danger of dying of intellectual starvation.

In the past few decades when women's books – and women's presses and bookshops – have appeared on the scene, books have once more assumed great importance among women. Again and again, women have testified that a particular book "changed my life"; and because they have found it so inspirational, they have bought multiple copies of it to give to (and change the lives of) their friends.

As the influence of the book starts to wane, what effects will this have on society? And on communication? How will it affect the position of women – and women's information, which is still in its infancy in some ways?

Along with the changes in the purchase of books, and the role that they play in individual lives, there has been a marked change in reading practices. It's not just that the next generation is decoding fewer lines of print; it's that they are decoding a great deal more visual imagery. They are reading screens. This is a different activity – physiologically, imaginatively and psychologically – from reading the symbols of the alphabet.

To sum it up: we are moving *from motto to logo!*

There's more to books than readers, of course, and authors are the subject of Chapter 4. Here too it can come as a surprise to find that authors as we have known them are a product of the printing press, and have only been around for a few hundred years. It's really only during the print period that we have developed the notion of the individual creative author as one who produces original material which is the intellectual property of the writer and represents work which must be paid for.

With the increasing importance of the Internet and the electronic media, the significance and the sales of books are threatened to the point where the death of the author is being widely proclaimed by certain members of the literary-theory community. Yet even as the print authors find the environment less hospitable, the multimedia creator (in which print is but one of the many media) is enjoying greater artistic opportunities.[15] Perhaps we are seeing just another stage in the evolution of authorship; perhaps this is another example of the creative use of new media, as happened in the past when we went from storytellers and minstrels to essayists and novelists.

What we do know is that more people today can publish their work to a bigger audience than ever before, and they can do it without the aid of middlemen. Assuming their connections to the Internet (a big assumption), authors now, with a stroke of a key, can reach millions, without having to go through the intermediaries of publishers, distributors, booksellers, etc.

This is why some people have described the new technologies as leading to the democratisation of authorship. With the aid of the new technologies, the masses can not only read: they can now write as well!

And as the book – and increased access to information and power – led to a social revolution, we can now begin to appreciate the extraordinary impact that the new technologies are having on our society. We are in the middle of a revolution, and because multimedia publication is now available to the masses we have to rethink the nature of creative authorship, along with the nature of the creative goods that authors produce. And how they will be paid for their work.

Will there be a special group of professional writers in cyberspace? What will their creative products look like? Will they be more like scriptwriters, games-makers, film producers, or code writers than novelists? Is this a gain or a loss?

Will we have, in the future, novel games as a major source of enlightenment and entertainment? A package of story-lines, scenes and characters, (provided by a professional) which we can then use or "play" with? Will we bring together the old skills of the creative writer with the new trend of doing and interacting, of designing our own combinations, of choosing our own endings?

Or will the creative writer be replaced by an intelligent agent?

We look back on the manuscript era as a period where intellectual curiosity and creativity was stifled. We describe that period as the Dark Ages: it would be fascinating

---

15. Government arts policy in Australia reflects this, with increasing amounts of money being made available to multimedia producers; and of course as the print-based publishing tapers off, multimedia and CD-ROM publishing is expanding.

to know how future generations will look back on print culture and its achievements.

Will the print author – and the novel, which emerged two centuries after the introduction of the printing press, be viewed as a limited, linear, inflexible and deterministic form that in its rigidity was more frustrating than satisfying for "users".

What will all this mean for our educational institutions? Chapter 5 on education is one of the key chapters. This is partly because education plays such a major role in our lives and our society, and partly because every facet of education must also undergo dramatic revolution.

What we have at the moment is an education system based on print. It has changed very little over the past few hundred years. Anyone from the eighteenth century would instantly recognise a classroom, a blackboard, a school, a teacher.

Yet what we will have in the next few years is an education system that is part of computer culture. It is not just the physical environment that will be transformed.

Whereas books have encouraged us to think in terms of a stable body of knowledge, a form of content that we can read, digest, learn, and know, computers dispose us to think differently – to be engaged in a constantly changing process where information is not stable or fixed. It might be quite sensible, for example, to ask, "what's in that book?" But it would be absurd to ask, "what's in that computer?" In a computer-based education system it will no longer be a matter of *knowing* information; it will be a matter of *doing* it. The essence will be making it as you go and changing it as you want to. And you will have intelligent agents to help you – and your teachers!

Every classroom will be transformed. Every student will be doing different things. Almost all teachers will need retraining to teach new and different competencies. Some schools are already well on the way to being computer-based, and·they help to suggest how fundamental and far-reaching the information revolution will be in education.

Universities will also have to change their purposes and their practices. Scholarship, knowledge, research and teaching are significantly different when done electronically. As more and more instruction and "knowledge-making" are done with the computer and on-line, the demands made of academics will bear little resemblance to the traditional requirements. If video and screen appearances are the rule, it could be that screen tests will be more important to the new breed of academics than PhDs have been.

Chapter 6 deals with libraries, for they too are experiencing enormous change. Established to provide access to information for those who could not afford books, public libraries performed a public service and promoted a more equal society. Whether they should continue in this tradition, and make available computer services to those who could not otherwise afford them, or whether they should become "user-pay" units, and join the information business are questions worth consideration.

Already libraries have moved from being the quiet havens of print to noisier and busier centres of technological activity. Older people for whom libraries have been something of a refuge are coming to terms with the fact that they can no longer go to the familiar catalogue, but have to use a terminal to find what they are looking for. They can be overwhelmed by the range of new on-line services.

Libraries are crucial agents in the information revolution, and they continue to play a part in shaping what we think and how we look for it. This is why their classification systems are so important. Leave out some individuals or some areas and they might never again exist in the electronic record; and for this reason we should be scrutinising the data systems that are being set up to form the mindset of the twenty-first century.

Within print culture, it has been more the exception than the rule to include women's contributions. Chapter 7, "Women, Power and Cyberspace", starts with the fact that women were left worse off after the invention of the printing press. Given the current "figures on men" in the new technologies, the issue is – will women once again be worse off after the latest information revolution ?

The evidence is not encouraging.

When it comes to cyberspace, men have the power. But it doesn't have to stay this way. And it won't. Not if women are convinced of the necessity – and the desirability – of becoming involved.

The fact is that the new technologies are the source of wealth and influence, and those who are in there will be the information-rich. Not that economic rationalist arguments and the profit motive should be the only incentives to take up the mouse.

Books, of course, did make individuals part of the information-rich, and it was the case that those who didn't have them (or couldn't read them) were among the information-poor; but one of the best reasons for reading a book has been that it is enjoyable: stimulating, entertaining and pleasurable!

The same goes for the new media. This is why women have to get in there. For the wealth, the power – and the pleasure of it all.

Nattering on the net is a satisfying, affirming and delightful pastime. Or it will be when women are full participants in shaping the system and the rules.

The glass ceiling may be preventing women from getting into the top levels of general management, but it is also preventing them from getting into cyberspace in appreciable numbers. Yet this is where the new communities are being formed; this is where the new human values are being forged. And 6 per cent of the space is just not enough for half the population. This is how marginalisation and oppression are too readily constructed.

Women are needed even at this stage to rewrite the road rules on the superhighway. Computer-competent women are needed to "suss out" this new public place and pass on advice to the next generation.

The last few decades have seen an explosion in women's knowledge, as women's studies have appeared on every campus and women's issues have become part of the public agenda. But if the present generation of women cannot hand on this knowledge that they have created, then the entire tradition is at risk – and it could be lost. This is the pattern of the past; it should not be the prediction for the next century.

That women have not been engaged in the new technologies to the same degree as men says more about the nature of the technologies than it does about women. Where women have made the technology accommodate their needs, their success knows no limitations.

Chapter 7, "Women, Power and Cyberspace", concludes with discussion of some of the women who natter so well on the net. They are models for the future. While cyberspace is a very different world from the one we know now, it will be even further transformed once women are full members of the cyber-community.

And this could be closer than you might think.

It is imperative that women become involved in the cyber-community in the same numbers and on the same terms as men. (It is just as important that Aboriginal and Torres Strait Islanders, and other marginalised groups, be fairly represented in the new medium.) There is no other way of ensuring that wisdom or knowledge can emerge from all the masses of information and data that are currently being produced.

In Western society, we now refer to the the manuscript culture of the Middle Ages as the "Dark Ages". We think of the people of that time as closed-minded and prejudiced, believing the most extraordinary and bigoted things about the meaning of life and the nature of the universe. There was plenty of information around at the time (including such things as the number of angels who can dance on the head of a pin), but because it was controlled by the powerful elite of the Church (who were interested in staying in power), we look back on the era as unenlightened; as a period without much knowledge or wisdom.

Then came the printing press. The history of the past five hundred years has been the extension of information-access and power to more and more members of the community. The Church gave way to the university as an information-producer; the priests of the Church were displaced by the priests of science, and scientific method helped to expand our understandings and our horizons.

The aristocracy made way for democracy, and more and more working people became literate and obtained access to education, information, and participation in the business of society.

The Western world called this progress. The more people who could be involved in information production and access, the greater the sum of human experience that could be included, and the more comprehensive the understandings that could be arrived at.

The last few decades have seen the extension of information production to women. In the 1960s and 1970s the women's protest was not just that women had been left out of society's database, but that in excluding half the population from information, the society ended up with only half-truths. Without the inclusion of women, the only information that could be produced was highly distorted.

The establishment of women's presses, the explosion in women's books and the introduction of women's studies courses in most universities around the world during the 1980s have helped to remedy the glaring omission of half the experiences of humanity from the repository of human information. The more women have put in over the past decades as information-makers, the more knowledgeable and wise the society is able to become. And this is why we simply cannot afford to permit white male dominance of the new communication technologies.

Not just because it's not good for the males. But because it's not good for society. A very distorted view of the world is created when only one social group, with one set of social experiences, pronounces on how it will be for all. This is Dark Age behaviour, rather than enlightenment.

In talking about the revolution that we are now experiencing, I am often confronted by people who insist that all we get from computers is information, in comparison to the great value-added dimension of wisdom which evidently flows from the book. While I have no difficulty in making a distinction between data and information on the one hand and the human component of interpretation and wisdom on the other, I cannot understand why the charge is levelled specifically at the computer. Print contains no more knowledge or wisdom than does the digitised image.

"It will be the end of knowledge and wisdom when we are all reduced to computers," I was told recently in no uncertain terms. My only response was that knowledge and wisdom lie not in the medium. I am sure I am not the only one to know many people who have read a helluva lot of books without any appreciable increase in terms of perspicacity, discernment, judgement or intellectuality. Whether it is oral, manuscript, print or electronic media, the issues are *access and equity* if we want the full story. We need to democratise information-production if we want the best possible basis for making up our minds and arranging our community.

Which is why this book is about people rather than computers. The discussion is about our intellectual growth and options, and the way we organise society in cyberspace – and in real life. The challenge for us is to transform information into wisdom in the new medium. It is to learn to live with computers – and to make a better world.

1

PRINT

Monks and Manuscripts

The printing press changed the course of human history. It produced an information revolution. It changed what human beings know, and how we think. This is why print is such a valuable starting point for understanding the views and values of our contemporary community. And in examining some of the changes that took place with the introduction of print, we can also see the parallels with the changes that are taking place with the current information revolution: it too is altering the course of human history.

Church Rules

Before print, there were manuscripts. They were the information medium of the Middle Ages. And it was no less of an *information revolution* when manuscripts were replaced by books, than it is today as books are being replaced by electronic media.

The high point of scribal (or manuscript) culture was the fifteenth century, just before the printing press made its appearance. This was a time when the Catholic Church ruled almost every aspect of existence in much of the European world.

The Church was not just in charge of life hereafter; it also controlled information on earth. It was in the powerful position of being able to control what was known, of being able to manipulate people's minds. Everything that everyone needed to understand was supposed to be recorded in the sacred manuscripts. These were written by hand[1] and almost always in the Church-controlled scriptoria,[2] the only exceptions being a few royal households where private collections were kept.

It was mainly monks who sat around in the scriptoria copying these ancient texts (although in the nunneries and convents there were also women scholars who were

---

1. Manuscript, from *manus*, Latin for hand.
2. *Scriptoria:* "writing rooms, specifically the rooms in a religious house set apart for the copying of manuscripts", *Oxford English Dictionary*.

scribes). The art of being a good copyist was to be able to copy absolutely accurately from the original biblical manuscripts, religious commentaries and sermons which were the knowledge, or information base, for the entire community.

Of course, no one was allowed to disagree with these official records that had been handed down through the ages and which were held by the Church authorities to contain the meaning of life and the secrets of the universe. So any copyist, for example, who changed anything, was not seen as creative – as making an original contribution – but was likely to be charged with *corrupting* the text.

And anyone who criticised the Church's view of the world (in which Church leaders were dominant) was also in for a rough ride.

According to the Church and the sacred manuscripts, God made the world in six days; man had dominion over all; and Adam came before Eve. And there would be no disagreement. What could be known was already known, and it was the role of the Church to guard it, to preserve it, and to keep it "pure" and free from any changes. Anyone who questioned or protested was quickly labelled as a heretic, and faced imprisonment, excommunication, or death at the stake.

Understandably, most people went along with the Church's version of creation and its explanations of world order and the state of things.

In 1450, prior to the advent of the printing press, the Church was able to keep control of information in the hands of a very few. It achieved this partly by conducting the business of the Church in Latin – a language which no one spoke as their native tongue, and which most people could not understand. Latin became a sort of secret code to which the Church held the key. (Until recently, medical doctors did the same thing, demanding Latin as an entry requirement to their education system, and using it for their prescriptions, etc. so that no lay person could know what was written. As it worked for the priests of the Middle Ages, so it worked for doctors of the twentieth century, to create mystique, and to add to their status and influence.)

In order to read the religious manuscripts, you had to know Latin: and in order to learn Latin, you had to enter the Church's education system, which was a training for the religious life. By such means you were initiated into the values and viability of the Church. With such a closed system, it was relatively easy for the Church to have an absolute monopoly on information and education.

There were no independent schools, no public libraries or state universities where anyone could go and develop an alternative view. There was just one official explanation for the way the world worked: just one word and one truth: just one system of authority.

The scholars – the custodians of the religious truths who had worked their way up through the Church's education system – spent their lives studying the sacred texts that had been handed down through the ages. They pored over the biblical stories, the religious commentaries, the sermons and the prayers that were the sum total of knowledge of the period. Scholars would work together, reading out loud, reciting the material. As the most respected authorities were those who could quote manuscripts from memory, much effort went into learning everything off by heart.

Most of the population – the laity – could neither write nor read. They were dependent upon the Church for information. Instruction was regularly delivered: each week, for example, the lesson was read from the pulpit.[3] There were also other ways that the Church issued its edicts.

There probably never has been a more monolithic system of information control than that of the Church in the Middle Ages. But even this absolute system came to an end once a new information medium made its appearance. The manuscript period, with its particular values about knowledge and power, all ended with the printing press.

Printing Presses

About 1450, printing presses began to appear all over Europe. The printers who started these presses rolling were a particular breed. As the inventors of these new machines, they had more in common with mechanics than they did with clerics; as entrepreneurs who wanted to use their presses to make a living, they were more like merchants than religious scribes.

Of course, these new printers were condemned outright by the priests.

This did not prevent them from churning out a lot of new publications which the Church had no authority over. It was from these various presses in the hands of "ignorant inventors", rather than from the old religious scriptoria, that information began to flow in the latter half of the fifteenth century. In a relatively short space of time, the Church's monopoly on knowledge was broken.

No matter what the Church did to stop this seditious information from spreading, the books kept rolling off the printing presses. Printers could be burned, books could be banned, but the Church continued to lose the battle for the control of the word. People began to know things that the Church did not teach. An alternative way of looking at the world was beginning to emerge. In this new perspective the Church was far from central; science, rather than religion, became the framework for explanations. The Church could be looked on from the outside. And it was found wanting.

This marked the beginning of the end of the Church's control of information. There was another way of explaining the world, apart from the religious version which represented the Church as all-knowing and all-powerful.

Reformation

The Church's loss of control was hastened by the Reformation and the actions of Martin Luther. He was the man who nailed his objections to many religious practices – in Latin – to the church door. He was by no means the first courageous cleric to protest, but he was one of the first whose cause was aided by the printing press.

Within days of his declaration against corrupt Church customs, Martin Luther's notice had been translated into local languages, and was published in thousands of

---

3. Reading the lesson is discussed in more detail on pp. 45–6.

copies. Anyone who could read (and you can imagine the increased pressure there was to learn to read when all this was happening) could follow the argument. They could appreciate what the protest was all about; they could even take the side of the Protestants.

The Church was caught in a bind. It could ignore at its peril the leaflets and posters which were circulating so widely and which were so critical of its practices. Or it could descend to the same vulgar level.

It decided to counter-attack. And so began the first poster war in history. The Church's critics leafleted the masses; and the Church tried to defend itself in a medium that it despised and condemned.

The winner was the printing press.

The Church was split into Protestants and Catholics. There were far-reaching consequences. While the Catholic Church survived, it was in a very different form. It lost its monopoly on information, on truth and meaning. It became a competing voice, rather than the only voice, in decreeing how the world works. And it too turned to print as its medium.

Generation Gap

The printing press itself put increased pressure on the population to learn to read. So many new books appeared so quickly, and on so many new topics, that there was a mad scramble among some people to cover everything. After the repetitive and conservative manuscript period, where the same old texts had gone round and round, the excitement of the new that came with the printing press must have been difficult to contain.

The changes which took place were indeed revolutionary. In the manuscript period, questions and speculations had not been allowed. All was known, and the purpose of scholarship and intellectual activity was to memorise it. And then came print and all the new publications which suggested that there was much that was unknown. Suddenly, it became possible to think that the Church did not have all the answers, and that questions, criticisms and speculations were the most stimulating (and not necessarily the most evil) intellectual activities.

In this climate, you can see how individuals and society changed dramatically as more information became available.

But it was the young who changed most, and who changed first. With so many books offering such wonderful opportunities to explore and extend understandings, it was the younger generation who were passionate about learning to read. They wanted to be part of the action and to have access to all these new ideas.

(This urge to be part of the new and powerful discourse has its parallels today with the great need to learn English – or American! As the language of international communication – of computers, broadcasting, air traffic, etc. – American English is necessary for participation; as with reading in the sixteenth century, you have to know the code to know what's going on. Language classes in English or American English are popular everywhere from China to Russia.)

After the relatively small number of manuscripts of the Middle Ages, the flow of books from the printing press must have looked like a knowledge explosion of incomprehensible dimensions. There is some debate about the figures, but the consensus is that by the year 1500, millions of books (perhaps even up to 20 million) had been published.[4]

Elizabeth Eisenstein tries to give some idea of the implications of this when she writes:

> A man [*sic*] born in 1453, the year of the fall of Constantinople, could look back from his fiftieth year on a lifetime in which about eight million books had been printed, more perhaps than all the scribes of Europe had produced since Constantine founded his city in AD 330.[5]

The sheer volume of books was not the only incentive for learning to read. It must have been enormously exciting during this time when the battle of the word was going on, and the Catholic Church and the Protestants were fighting it out in print. To have been unable to read would have meant being left out of the most important events of the period.

This is one reason that reading got a bad name. It was the way to radicalism. Those who could read were not under the Church's influence. Not only could they keep up with the threats to the authority of the Church, they could also find out things for themselves – from books.

The Church's hold over education, over the minds and manners of young people, was undermined, as the young discovered they no longer needed the religious establishment. They could learn to read, and then be responsible for their own education. They didn't have to become part of the Church to have access to information.

An enormous generation gap developed. The young learnt to decode the new medium; they took to books and went off on an intellectual adventure of their own where their elders did not follow. Instead, the older generation tried to prop up the authority of the manuscripts and the old ways of studying and learning. The gulf between manuscript and print – and the Church and the laity – widened.

As print gives way to the electronic culture at the end of the twentieth century, we can see the similarities. The young are much more at ease with the new medium: the average five-year-old can program the video better than the average fifty-year-old.

---

4. The contemporary critic, Alvin Kernan, states: "The print revolution began in the mid-1400s and print began to affect culture at once, producing, in place of the few books that the scriptoria produced . . . between ten and fifteen thousand titles, or at the minimal runs of five hundred copies, up to 7.5 million books in the fifty years after 1450." Alvin Kernan, 1990, *The Death of Literature*, Yale University Press, p. 129. Historians of the book, Lucien Febvre and Henri-Jean Martin, go further: "Assuming an average print run no greater than five hundred, then about twenty million books were printed before 1500." Lucien Febvre & Henri-Jean Martin, 1984, *The Coming of the Book*, Verso, London, p. 248.
5. Elizabeth Eisenstein, 1983, *The Printing Revolution in Early Modern Europe*, Cambridge University Press, pp. 14–16.

The End of the Old Order

The world changed when the primary information medium changed from manuscript to print. The story of creation, the meaning of life, the notions of good and evil – in other words, *the nature of knowledge* – were transformed, as print swamped the population with new ideas, and challenged the old forms and explanations.

Along with the changes in knowledge went changes in social and political organisation. Job restructuring took place on a grand scale, just as is the case today as print itself is being replaced by electronic media.

From the early sixteenth century, one of the first job casualties was that of the copyists. The many monks and nuns who had worked on all those beautifully crafted and illuminated manuscripts – who had been respected and revered for their marvellous talents – suddenly found that their precious skills were no longer wanted. When information could be reproduced so quickly, and so accurately, by the printing press, what place could there be for copyists who took so long to produce so little, and so inaccurately in comparison?

The copyists and scribes had been born into a world where manuscripts had been handed down through the centuries, and these texts were held to be the repositories of human knowledge, the word from the mouth of God. It was inconceivable to them that these sacred texts could become irrelevant and obsolete.

What about all that ornate work, that copperplate writing, those superb illustrations? What about the very essence of a manuscript which was the accumulation of the sacred sources, along with all the many devoted hands that had crafted it over the years? It was beyond the comprehension of the copyists that these documents, and the skills that went into making them, should just be cast aside.

In much the same way, it is inconceivable to many people today that the book could become marginalised.

We have come to attach many of the same values to the book that were once associated with manuscripts; we have come to see them as a symbol of wisdom and artistic achievement, and we cannot readily accept that their role and their representation is changing. It took a long time – centuries in fact – before the book took on some of the mystique that once was associated with the manuscript. But now that we have constructed this role for the book, we are defensive about any changes to it.

Yet it is possible, if not likely, that future generations will attach the same value and reverence to the electronic media, particularly the computer, though most print-reared people cannot be convinced of the likelihood of this sacrilege.

On any level, the passing of the book (and of the author and literature)[6] has to be a loss. To those who have been avid writers and readers, whose lives have been constructed from books and print, the demise of the book actually means the end of a form of personal existence. It will be the finish of certain meanings, habits, memories

---

6. For further discussion, see Chapter 2 on literature and Chapter 4 on authors.

and pastimes. The fact that the book may be an object of beauty in its own right, and that it may contain some of the secrets of life, will not be sufficient to protect it from progress. No one could dispute that the manuscripts of the Middle Ages were truly beautiful objects, or that they enshrined much of human experience and wisdom, but that didn't stop them from becoming museum pieces.

While this was clearly a loss – a death, if you use today's terms – and a matter of regret, the end of the manuscript also meant the rise of the book. It made possible the emergence of fiction, the growth of a reading public, the empowerment of millions of people, the development of print culture. And this can't have been such a bad thing.

No doubt the same process will operate with the electronic media.

Future generations will be able to look back and evaluate the book in the way that we look at manuscripts. They will agree that it was a wonderful medium, a marvellous source of information, entertainment and illumination, and that there were many who were sad to see it go as the primary medium.

But it made way for the computer and the television – for the electronic culture and all its achievements.

That day, however, is quite a way hence. In the meantime, many of the defences that are being used now to maintain the importance and status of print are very like those that were used in the seventeenth and eighteenth centuries to maintain the importance and status of the manuscript in the face of the print threat. The medium may change, but human responses seem to stay much the same.

The most common objection to the book that was put forward by the manuscript supporters was that it *lowered the standards*. All sorts of standards.

To begin with, there were the aesthetic considerations. The beauty of a handwritten manuscript could not compare to the machine-produced, alienating, soulless nature of print, the copyists disdainfully declared. Their objection to the debasement of print didn't stop with the product; they were equally damning about the process.

It has to be noted that the scribes and scholars had studied collectively; they had spent their days meticulously making their way through a manuscript, reflecting on every word to fix its meaning. They discussed and debated intensely. They read to each other and recited from memory. They knew their manuscripts off by heart. This was what it meant to be a scholar, to be learned and respected. To them, this was what reading was all about: it was being a "knower" of religious texts, of the word.

Reading was a highly specialised activity, entrusted only to an elite few who had been carefully trained by the Church in their responsibilities. Reading/studying was an art that only the specially talented could expect to accomplish, and which took years to acquire; it had to be undertaken in a spirit of great seriousness and reverence. Books, and reading, mean something entirely different now.

The book was portable and replicable. Readers could go off on their own, to read (or scan or zap) to themselves in private and to make up their own minds. The fact that they could race through one book after another was yet another indication of the way standards were falling, in the eyes of the scholars and scribes. "No one could process that much information so quickly," they declared in tones that have their contemporary echoes.

As the book increased in popularity and more and more people outside the church took up reading (thereby deskilling the old authorities), the predictions of gloom and doom from those who had most to lose became even louder.

According to the Church, the new custom of allowing everyone to go off and read in isolation and to develop opinions of their own was a recipe for anarchy and disaster. It would lead to a breakdown in communication, and the emergence of a society of individuals – a state which, at that time, was not considered at all desirable.

There wasn't any aspect of society that the Church didn't think was being ruined by the dreadful medium of print: the historian Raymond Williams quotes examples from the seventeenth century in which allegations were made that the book would mean the end of conversation in the family.[7]

The members of the establishment who were so concerned to discredit the book and who protested so vehemently about the drop in standards, were simply trying to protect their own power base; they were trying to hang on to their own skills – as skills. They wanted a world where what they could do was valued; they didn't want changes which marginalised them, deskilled them, and left them feeling worthless and useless. That is why they kept insisting that the old ways were the only ways and that the new should be fiercely resisted.

While these issues are explored in more detail throughout this book, the point that needs to be made here is that there was a time when reading was considered just as damaging and dangerous for some sections of society as television (and computer games) are considered corrupting today.

Writing about the seventeenth century, the great contemporary literary scholar, Alvin Kernan, says that "Reading came to be feared in much the same way that too much television viewing in late twentieth century America has become a kind of cultural bogey."[8] As with many complaints about television viewing, reading was once seen as a great threat to public health; the couch potato, however, seems to suffer from a slightly different range of risks and ravages from those that faced the bookworm.

In 1795 the physical consequences of excessive reading were outlined as follows: "susceptibility to colds, headaches, weakening of the eyes, heat rashes, gout, arthritis, hemorrhoids, asthma, apoplexy, pulmonary disease, indigestion, blocking of the bowels, nervous disorder, migraines, epilepsy, hypochondria and melancholy".[9] It's clear that access to the new sources of information wasn't encouraged by those who were the masters of the manuscript era. But if they didn't stop the manuscript giving way to the book, they do provide interesting ideas and insights into an earlier information revolution.

---

7. See Raymond Williams, 1978, *The Long Revolution*, Penguin, Harmondsworth, Mddx, for further discussion.
8. Kernan, *The Death of Literature*, p. 130.
9. Robert Darnton, 1986, "First Steps towards a History of Reading", *Australian Journal of French Studies*, 22, p. 5.

Professional, White, Male Standard

If there is one thing that print has given us, it is the concept of standardisation. Partly because print itself doesn't change, the medium has helped to promote a mindset in which we want other aspects of life – and language – to remain fixed and unalterable.

Print has allowed us to develop static and unchanging standards and benchmarks. These have been of enormous assistance in promoting the scientific and industrial revolutions, but they have distorted and damaged many of our understandings and practices about language and its use.

Linguistic Change

One thing we know about language is that it keeps changing. Expressions keep changing: the meanings of words keep changing: pronunciations keep changing. In English, you only have to go back to earlier versions of the Bible or Shakespeare to realise how many changes take place in the language as each generation bends it to accommodate their own needs and purposes.

In contrast, one of the most salient features of the print medium is that once words are in print, they remain unchanged. They are a permanent record. The only way the words can change is if the materials on which they are printed deteriorate; they can disappear as the paper disintegrates.

The ability of the printed word to remain constant has been both a burden and a blessing. It has the advantage of being able to stand as a record, to be preserved; it has the disadvantage of capturing the language at a specific stage and making it still, even though language in use is a dynamic system. This is why print has the dangerous capacity to divorce language from the people who use it; it fixes the language at a precise point, when the people who use it keep changing it as they go.

The gap between print and language use probably wouldn't be such a contentious issue if it were widely understood that the language keeps changing. What was good form in one generation becomes undesirable in the next, and vice versa; words, phrases, usages that were once unacceptable are now widely used. The test of the pudding is in the eating; if a significant number of people use them and are understood, then the new usages are acceptable usages. That's the only way the words got into the language in the first place!

But the notion of a changing language in which the community is seen as the creator of the symbols and the meanings (and where communication is what it is all about) is almost alien in the twilight of the print period.

The blame for this can be laid partly at the door of educational institutions; for years and years they have taught fixed (and false) rules about language to generations of students. Without this instruction, students would not necessarily come to think in terms of rules and regulations and correct and incorrect usage. After all, people had no trouble acquiring a language and a form of expression before the arrival of the printing press and the grammarians who made up the standards; neither Shakespeare nor the early feminist writer Aphra Behn suffered language difficulties because the rules of correct English had not in their days been invented.

The link between fixing print and fixing the rules about language is not hard to make. But these rules and regulations are not language; they are a form of etiquette. They are not the food of communication which sustains the relationships of a language community; they are linguistic table manners. They are all very well if they are designed to make things more pleasant and promote harmony. But unfortunately, many of the linguistic table manners that were devised during the print period are more tests of who is in and who is out, than courtesy.

A standard has been set within print culture. It has been the standard of white, professional men that has become enshrined as the way things are done properly in print. This white, professional, male standard has been there for all to see. It permeates everything: accent, spelling and grammar, norms of expression, vocabulary and "meaning". The dismay and distress at the passing of the print era has more to do with bringing to an end a patriarchal presence that has been encoded in communication than it has to do with the loss of print.

People who have not read a book for years can be vehement in their defence of print values and in their commitment to preserve the white, professional, male perspective that is at its centre. It's not the print so much as the point of view that is under threat.

Print as Distortion

Because language is dynamic and print is static, there is always a gap between the spoken and the written form.

This is one of the reasons that some people were against "putting it in writing" from the time that writing was invented. The Greek philosopher Socrates, for example, was very critical. He objected to the way ideas were fixed once they were "on paper". He insisted that this process was the very opposite of thinking. Intellectual activity called for an exchange of ideas, he claimed, with feedback constantly modifying the contributions of the speakers. He was adamant that the written form could not cater for this interaction. You could only follow an argument once it was in writing, he declared: you couldn't engage with it, or change it. And the changing was the essence of ideas and understandings – and communication.

Socrates himself would not put anything in writing: he predicted that ideas would be distorted by the very act of being pinned down, and he wouldn't allow his own thoughts to be misrepresented in this manner. (Fortunately, Plato had no such objections: he recorded all the criticisms that Socrates listed.) If he could see what had happened during the print period, there's no doubt that Socrates would feel that his protests had been more than justified. With its ability to fix the language, and the ideas, over the centuries, print has limited and skewed the active process of thinking and talking.

"Print has changed the way we communicate," states Hugh Mackay; "By forcing us into the disciplined mould of a logical, linear, rational medium, print encouraged us to accept logic and rationality as inherent virtues."[10] There *are* other ways of looking

---

10. Hugh Mackay, 1992 "Watched Any Good Books Lately?", *Bulletin*, 11 August, p. 45.

at the world; print has served us well in a law-governed and rational era – which the scientific age has been. But the electronic media offer us an entirely new, more dynamic and relative point of view, which has both possibilities and limitations.

In looking back on print culture, the consensus will be that it was the period of standardisation. It gave rise to a mindset of regularity and repeatability (the basis of scientific evidence). The main virtue of print was its uniformity and repeatability; the printing press could produce thousands and thousands of copies of a text which were *all exactly the same!* Standardisation flourished.

This was in stark contrast to the handwritten manuscripts that until then had been available.

Standardisation Mania

Once uniformity became possible, and the order of the day, standardisation was not only understandable: it was desirable and useful when it was first introduced:

> booksellers were primarily concerned to make a profit and to sell their products, and consequently they sought out first and foremost those works which were of interest to the largest possible number of their contemporaries. Hence the introduction of printing was in this respect a stage on the road to our present society of mass consumption and of standardisation.[11]

This capacity to produce identical and repeatable copies of a text had enormous advantages. It was this element of uniformity and accuracy that allowed the printing of graphs, charts, maps and calendars which were all exactly the same. This was a necessary condition for the development of mathematics (which made use of tables, graphs, etc.) and even for music, where a uniform system of notation could be produced as often as required.

But uniformity also had its price. Standardisation soon became an end in itself: instead of being an aid to communication, it could be a set of rules and regulations that were imposed on the language. An additional system developed that speakers, and readers and writers had to learn (by heart!) if they wanted to show that they knew what was the standard form, what was correct; otherwise they would be caught out making "mistakes" and using the wrong "language fork".

As print slips from being the primary medium of information, many of these features of standardisation are no longer as important. In an image-based society, standardisation or correctness of the word doesn't carry the weight that it used to. But this is not a drop in standards, as some people would like to insist; it is a change in medium. It could even be an improvement. We are seeing the removal of a set of rules that came to have more to do with setting up a pecking order than with helping communication.

From the time that the first sheets rolled off the printing presses, there was a compulsion to keep things in order. To produce linguistic uniformity. It even extended

---

11. Febvre & Martin, *The Coming of the Book*, p. 260.

to the typeface that should become the standard for books. With the passing of the scribes, and their variegated hands, "styles of lettering became sharply polarised into two distinct groups of type fonts" says Elizabeth Eisenstein.[12] Would the standard be Gothic or Roman?

That so many people today want so desperately to hang on to this uniformity, to the old and arbitrary standards, raises the issue of what role these rules play in influencing world views, in setting human horizons. It is easy to demonstrate that most of the rules about correct and incorrect ways of using and representing the language are nothing other than the particular prides and prejudices of powerfully placed white men of the past.

Yet people who embrace changes in many other areas, who seek more just legislation or more equitable social policies, can become inchoate with anger when confronted with the evidence that grammatical rules and spelling conventions have little to do with language and a lot to do with creating an unequal class system. They don't want to let go of language rules that have been devised to keep people in their place. The issue of correct spelling, for example, can bring out the worst in otherwise liberal and fair-minded individuals.

Spelling It Out

A Sound System

Prior to the appearance of the printing press, people were perfectly able to spell. It's just that they spelt each word according to the way it sounded. This had the advantage of allowing spelling to keep pace with any changes that took place in pronunciation. The different spellings could also reflect the different pronunciations in regions and dialects.

In some ways it could be argued that spelling then was a more creative and intellectual activity. You had to *think* about how you would spell a word, rather than having to recall or remember the standardised version, which was later to become the hallmark of print.

There were many variations in spelling even after the printing press, but they don't seem to have caused the population undue stress. Shakespeare, for example, is supposed to have spelt his name sixteen different ways, yet this creativity on his part doesn't seem to have led to any confusion. No one presumed that the man was illiterate, or of weak character, because he could come up with so many ways of spelling his own name.

This was because the notion that there is only one way to spell a word – a fixed, standard and correct way – had not really taken hold in Shakespeare's day. The spoken and the written forms both continued to change. It wasn't until the nineteenth century

---

12. Eisenstein, *The Printing Revolution*, pp. 54–5.

that standard spelling (and of course non-standard, which went with it) became entrenched. After that, it was possible to make judgements about whether someone was ignorant – or immoral – if they didn't spell a word according to the rules. Spelling conventions came to be used as a way of putting people in their place. This is a long way from the art of communication.

To start with, the standardisation of spelling had some advantages: "When they took over from the scribes, printers began to eliminate the whims of spelling and the phrases of dialect which would have made their books less readily understood by a wide public," comment Lucien Febvre and Henri-Jean Martin.[13] The fault was in keeping the *same* standard spellings when the pronunciation changed. All could have been well if spelling too had been constantly updated to keep pace with language developments.[14]

But it wasn't just this gap between letters and pronunciation that made for difficulties. There were other forces at work which helped to make the distance between the two even greater.

Once spelling became fixed, "it came to correspond less and less with pronunciation," write Lucien Febvre and Henri-Jean Martin, and the whole process was increasingly "complicated by the influence exercised by classical languages".[15] As the classics were translated from Latin and Greek, into English (often by way of Spanish or French), a great number of foreign words, which were not pronounced at all in the way that the letters suggested, were introduced into the English language: this made the task of spelling "correctly" even more capricious.

The Right Way

Correct spelling has been held up as a measure of literacy (or illiteracy). Those who have taken the same licence as Shakespeare and spelt their name in a variety of ways have been labelled as non-standard, or as character deficient.

That is not the only side effect that there has been: countless members of society have had their confidence eroded – and their writing and reading efforts dampened – by the trap of correct spelling. It isn't too much to suggest that the obsession with standardisation (and the punishment of non-standard or "incorrect" forms) has contributed to the ratings on illiteracy. With so many mistakes to be made, some people are just too frightened to try to write or read. Yet these "mistakes" didn't even exist in Aphra Behn's or Shakespeare's era.

Spelling has definitely been used throughout the last century of print to make moral judgements about individuals and the state of society. It was not uncommon – prior to spellcheckers – to find that job applicants would be eliminated on the grounds of a

---

13. Febvre & Martin, *The Coming of the Book*, p. 320.
14. To some extent this is what the Americans tried to do – to rationalise spelling and pronunciation. But for it to work, they would have to keep on doing it.
15. Febvre & Martin, *The Coming of the Book*, p. 319.

spelling error, for example. To be unable to spell was a measure of ignorance; to neglect to look up a dictionary and find the correct form was considered a measure of carelessness, slothfulness and irresponsibility – all of which were disqualifications for the job.

The conventional wisdom now is that young people don't spell as well as their parents; that computers and spellcheckers mean that they don't keep the rules in their heads any more, but are content to access a computer program. This is fine, as far as most young people are concerned; they want to be able to *do* it, not to *know* it. But it is often taken as a sign of moral decadence by the older generation, rather than as evidence that the information medium is changing.

People spelt creatively before print, and no doubt they will again after the values and mindset of standardisation have begun to recede. It could be that in the not too distant future, when the population once again spells a word the way it sounds (as they work on their e-mail or electronic bulletin boards), that scholars will look back on the print period, and standardisation, as the aberration in communication. Then they will see mindlessness not in those who defied the spelling conventions, but in those who insisted that such arbitrary mechanics should ever have been committed to memory, and then used as a standard of worthiness. (This is not to suggest that we couldn't do with some consensus about the form, given all the variations of English/American pronunciation from one country to another.)

Use Your Own Spell Check!

There are many reasons that standardisation in spelling is not as important today; it's partly because of spellcheckers that the same energy is not put into drilling the young to memorise all the rules. They simply don't need to learn them off by heart any more, if they are working on a computer. But there is also a new flexibility associated with the electronic media which is at odds with the very idea of standardisation and regulation.

Even where print appears on the screen, it is not fixed; it moves, it can be changed. There isn't the same sense of a finished product, or of a fixed meaning. Electronic "publications" have an air of fluidity, fleetingness, ephemeralness about them. Pluralism and diversity are much more the current values, than standardisation.

So, for example, when I come up with my own creative representations of words on my e-mail, those whom I am contacting don't think that I am ignorant or reprehensible. Neither the medium nor the form is seen as definitive or authoritative as the printed page used to be. Those who are connecting to my messages are likely to assume that I am not a good typist, or that there is no spellchecker on my e-mail – if they notice my non-standard offerings at all.

I have no doubt that they will not only work out what my message means, but that I will provide them with some entertainment in the process.

Besides, if there is anyone who does object, they have the solution close at hand. Let them use their own spellchecker to standardise the message. But it is their problem, not mine, if they cannot cope with the pluralism.

The real ignorance (and perhaps even immorality) is that of confusing the desire for standardisation and pecking orders with the way that language works.

The Glory of Grammar

Making up uniform rules didn't stop with spelling; it extended to grammar as well. It had much the same rationale – and much the same results.

The Pride of Grammarians

"The standardisation of grammar and vocabulary is even more important than spelling in establishing a language that can be readily understood by one and all," write Lucien Febvre and Henri-Jean Martin. This is why Martin Luther, who was interested in reaching as many people as possible, "made efforts to rid himself of his own native dialect".[16] Printing presses had only just started to publish in local languages, such as German and English; this was before there was any great sense of a national language, so there were many regional variations. Martin Luther tried to find the most common expressions, that is, the ones that the greatest number of people would understand.

Even if the initial moves for standardisation of grammar were like Martin Luther's – in the interest of extending communication – they soon took on other functions which had little to do with helping people understand one another. But, like spelling, they had a great deal to do with signalling who was in the know and who was not.

Language is a powerful tool for establishing hierarchies. It also influences what we think, and what we can see and understand. One thing can be identified in the making of print culture: those who had the power to make up rules about language generally did so on the basis of their own self-interest. Unlike Martin Luther, who tried to work out what ordinary people were saying and, who in the interest of communication, then tried to use similar language, there have been many English grammarians who decided that the rule for language would be: what they did was how it should be done.

They were not necessarily motivated by the need for communication, but were more concerned with their own position and preferences. Their view was that their code was the best, the correct, the standard code – and that the ordinary people should follow their example!

One of the things that needs to be sorted out here is the role of grammarians. It's self-evident that grammarians don't come first. It simply isn't the case that there was once a group of grammarians who made up rules about the language, that the population then came along and learnt them and that, as a result, everyone was able to speak "proper" English.

Grammarians come into being *after* people have developed a language. You can see this with young children who are learning to speak. You don't teach babies

16. Febvre & Martin, *The Coming of the Book*, p. 322.

grammatical rules, and then wait for them to talk. You teach them to talk, and then later on someone can come along and deduce or describe what rules are being used.

This is how a descriptive grammar could be constructed.

But during the seventeenth and eighteenth centuries, quite another breed of grammarian emerged. These were not descriptive but prescriptive grammarians. They didn't try to describe what English speakers were doing: they prescribed what they should do.

In the mania for making rules which went with print, they laid down the law of correct grammar – which is, of course, quite arse about.

Most of these new prescriptive grammarians were scholars. They were well educated by their own terms. They had learnt Latin, which they persisted in regarding as the sign of an educated man. And, of course, almost all of them were male.

Many of these grammarians genuinely believed that Latin was a superior language and that English was indeed the vulgar tongue. They saw Latin as refined, sophisticated, the medium of a privileged and polished elite: English, in contrast, was spoken by the masses, and was crude, coarse and common. These same grammarians sincerely believed that the English language could be vastly improved with the addition of a little Latin finesse.

So they started to lay down the rules for English, not on the basis of the way people were talking (which Martin Luther had followed), but on the assumption that the language would be better if it were more like Latin.

Latin Lovers

It was important to the prescriptive grammarians that the masses did not speak Latin. They knew that the introduction of Latin forms would show the common people up as ignorant, as being of a different order. This, of course, was the point.

No grammarians seriously argued for the introduction of Latin rules on the grounds that they would help communication; that they would aid understanding between one group of citizens and another. All the justifications for Latin additions were that they would add refinement, distinction and dignity – and marks of class – to the otherwise vulgar English tongue.

So this powerful elite who knew their Latin imposed their prejudices on the language. For centuries, English speakers had been finishing their sentences with prepositions without any ill effects for the population or for communication. But the prescriptive grammarians suddenly declared that this usage was out!

As you couldn't end a sentence in Latin with a preposition, they insisted, you weren't going to be allowed to do it any more in English. Anyone who continued to use the vulgar form – who said that the new rules weren't on – would be branded as common and ignorant.

This is why for centuries students have been drilled in the rule that you can't end a sentence with a preposition. Not because the English language doesn't have such a construction: not because people would not be able to understand such usage. But because a few jumped-up grammarians of the seventeenth century happened to think that the Latin grammar they knew was superior to that of English. Print culture helped to enshrine their view.

It doesn't seem much of a reason for keeping such rules today. Or for thinking that the world is coming to an end if speakers now indicate that these are rules they don't want to put up with.

The plain fact of the matter is that the only basis for these rules in the first place was the snobbery of a few supposedly erudite men who wanted to put their Latin stamp upon the language. They wanted to keep some of the elements of an exclusive code. The only basis for retaining some of these rules today is the same snobbery.

This isn't to say that good communication isn't to be encouraged. It should be. That's the point. But communication isn't assisted one iota by not ending sentences with prepositions, or by insisting on "It is I" rather than "It is me". Such requirements have nothing to do with communication. Their purpose is to establish a linguistic hierarchy, to bestow a sign of membership of an exclusive club.

Linguistic Table Manners

As a product of print culture and all its rules, I have learnt all the laws laid down by the prescriptive grammarians who reigned throughout the print period. To this day, I notice if someone says "If I was" rather than "If I were". But I realise that this is *my* problem. That I perfectly well understand what the speaker means when she says "If I was", and the fact that she doesn't use the subjunctive after the conditional (as Latin speakers did thousands of years ago) doesn't matter a damn. (Unless, of course I am looking for the membership badges of those who were brought up in the same "superior" school as the one I was initiated into in my days at the University of Sydney.)

I was reared on the rule that you couldn't split an infinitive in English because there was no such construction in Latin. My liberation has been to really try to specifically split as many infinitives as possible. I must admit that it is galling to have an editor "correct" my grammar and change my split infinitives back to the "right" prescriptive form.

Sometimes it is almost beyond me to explain that I do know the rule, that I can *use* it, but that I don't agree with the pretensions of the eighteenth-century grammarians and that I am deliberately trying to subvert their snobbish practices. But sometimes, trying to point out to an editor that such a rule is increasingly a hangover from the standardisation mindset of print, and that neither then nor now do we need such impositions in communication, is beyond me.

It can be an uphill battle with editors and publishers (and some commentators and reviewers) who slavishly cling to the rules they were taught in the print period, and smugly declare that there is something wrong with those who do not follow these rules.

If there is one principle that is at the core of the new electronic media, it is that of pluralism: there is more than one way to print, to spell, to pronounce, and to talk. The idea of "correctness" in grammar is just as silly, and set, and snobbish, as that of correct spelling. Even a phrase like "How are youse going" (addressed to more than one person, of course) conveys meaning and communicates perfectly. So it can't be wrong. It can be different from the seventeenth-century prescriptions, but it can't be "incorrect"; it's just another more contemporary, more common usage. That's all it is. It's not a drop in standards, but a mark of the progress of the years.

The trouble with the concept of standardisation is that as soon as it has been set up – as soon as any one single model becomes the norm – it creates a lot of non-standard users. The ones who have not been included are generally seen as inferior and deviant in comparison to the benchmark. This is one of the reasons that women have fared so badly over the past five hundred years in a climate of standardisation.

The standard human being has been male. The position of woman as non-standard, as other and in the wrong (the way that Simone de Beauvoir described),[17] was intentionally constructed in the English language. It is a fascinating – and frustrating – exercise to follow the imposition of this form of linguistic male supremacy. There can be no more self-interested or blatant example of this process at work, than in the case of he/man. Men decided that *man* would embrace *woman*, and that the standard for humanity would be male.

He/Man Hoax

Man Before Woman

The first records that we have appear about 1550, one hundred years after the printing presses, when standardisation as an ideology was beginning to have its supporters.[18] The emerging school of prescriptive grammarians started to set down the rules during this period. One that they came up with was that it was more "natural" for man to come before *woman*, and that this should be given substance in the language.

This meant that it was more "natural", for example, to say male and female, husband and wife, brother and sister etc. (This is why I deliberately try to vary my own usage, so that at least half the time I say female and male, daughter and son, etc., just to eliminate the male supremacist usage that was imposed in the past.)

The structure of the language influences the way we see the world, and there can be no doubt that the principle of putting men first in the language had its repercussions in the society. The male grammarians were able to make some of their ideas about their own importance come true when they gave males pride of place in language, on the grounds that they had pride of place in the natural order.

As there were no females among them – no women students, scholars, or policy-makers – there were no voices raised against this one-sided argument. There were no

---

17. See Simone de Beauvoir, 1972, *The Second Sex*, Penguin, Harmondsworth, Mddx, p. 15, for discussion on women, by definition, being in the wrong.
18. For further discussion, see Dale Spender, 1985, *Man Made Language*, Routledge & Kegan Paul, London, pp. 147–51.

women who could testify that they didn't feel as though they were followers, and that they wanted proof of this natural order which would put them last, before they could agree with the "logic".

By 1646, another hundred years later, the position had been consolidated and the arguments started to appear that it was not just *natural* for men to come first, it was *proper* as well.

Then in 1746, the rule that males were superior was made even clearer. Rule Number Twenty-one in Richard Kirkby's *Eighty-eight Grammatical Rules* insisted that men were entitled to be more important because they were more *comprehensive*. The argument that the men put up was that *man* stood for much more than *woman*: that is, that there was much more to *man*, that he represented more of humanity. And for this reason *man* should become the standard form.

Apart from the fact that he was male – and that he had inflated ideas about his own importance – Richard Kirkby doesn't appear to have had any other qualifications for making his new rule: the only basis for *man* becoming the standard seems to be that he – and a few other like-minded men – felt that their sex had a greater claim to humanity.

Again, there were no reasonable, rational women in his circle[19] who might have told Richard Kirkby not to be so silly.

Making *man* stand for *woman*, however, was not the end of the matter. Many members of the community didn't take to the new rule, and it is not hard to understand why. It was obvious that more than half the population was not male, and to refer to these people as male was most confusing. To females, it all seemed such a nonsense. (According to research, women have used *man* to mean *woman* much, much less frequently than men, right through to the present day: because most know now, as they did then, that *man* does not include them!)[20]

Despite the best efforts of the grammarians to make everyone use the male term, some people were perverse and went on saying gender-neutral things like "Everyone should have *their* say" when the grammarians insisted that it should be "Everyone should have *his* say", thereby ensuring male visibility.

But the grammarians were determined that their rule would prevail. They worked to have it gain the force of law. Almost one hundred years after Richard Kirkby's rule appeared, an Act of Parliament was passed in Britain in 1850 by an all-male House: it decreed that *man* should be the standard form, and that by law it would encompass *woman*.[21] So does standardisation reach the heights of absurdity.

---

19. Elizabeth Elstob, 1683–1765, was one of the few exceptions, in that she was a woman scholar with an interest in grammar, but it was considered "unnatural" for a woman to be interested in or informed about such matters.
20. Wendy Martyna, 1983, "Beyond the He/Man Approach", in Barrie Thorne, Cheris Kramarae & Nancy Henley (eds), *Language, Gender and Society*, Rowley, Mass., pp. 25–37.
21. See Ann Bodine, 1975, "Androcentrism in Prescriptive Grammar", *Language in Society*, 4, 2, pp. 129–56.

By these means the male grammarians (and politicians) made themselves the standard human beings and imprinted this value in the community consciousness. In print culture, where standards and regulations have been part of the mindset, where the emphasis has been on getting things right and putting them in order, it was the image of the male that was fixed as the one correct, proper representative of humanity.

When I read of prominent women who want to be called *chairman*, I find their reasons embarrassing and upsetting. They are undoubtedly highly intelligent and reasonable women, so my assumption is that they don't know anything about the history of the language. Surely no self-respecting women would persist with the term that was quite deliberately chosen by men to advertise women's supposed inferiority and real exclusion?

Once women are aware of the reasons for *he/man* and know how it has been used against women, they will cease to be duped into thinking that there might be something in it for them to be called *man*. And once the standardised mindset makes way for the pluralist world view that goes with the electronic media, we will find it easier to have both women *and* men as the rightful representatives of humanity.

Of course, what we could need is more information in our education system about the way language works. It isn't an accident that most of the issues raised here have not been taught. They would have undermined the status quo. They would have challenged male supremacy, and race and class hierarchies as well.

But if we are going to educate a community for the twenty-first century, we could do with less on the arbitrary rules and standards of spelling and grammar, and more about power, people, and process.

Man Means Man

Women, much more than men, have been drilled in speaking properly and politely. While there are a number of reasons for this, no doubt one has been that for women, many of the rules and standardised practices are simply nonsense. They have to learn off by heart the "correct" forms and explanations, as they have no way of working out for themselves what is right and what is not. (Women would not be able to devise a test that would let them know when they were women and when they were men, for example.)

But ever since men started to argue that *man* stands for *woman*, it has been obvious to anyone who hasn't just parroted the rule, that this is a false argument. One of the popular pastimes of feminist linguists in recent years has been to catch men out on this claim.

There are countless examples which show that prominent male scholars think *man* means *man* as the standard form for human beings, even as they defiantly declare that the word is gender-neutral and that *man* includes *woman*.

So, in writing about the characteristics of mankind, Eric Fromm can state that man's vital interests are life, food and *access to females* – which gives his game away. Loren Eisley can say of man as a species that "his back aches, he ruptures easily, his women have difficulties in childbirth"; which makes it pretty clear that *man* means *male* and that he "owns" woman![22]

Adopting a slightly different line, Casey Miller and Kate Swift show how ridiculous it becomes if we do take man at his word, and include a female image in the "man as humanity" term: "One may be saddened but not surprised at the statement 'man is the only primate that commits rape'," they explain, because even though *man* refers to only half of mankind, there is no conflict between man as a species, and man as the agent of this act; Man is man. "But 'man being a mammal breastfeeds his young' is taken as a joke":[23] and it shows just how far removed from the meaning of man, as well as mankind, woman is.

Man no more means *woman* than *dog* means *cat* and anyone who keeps on arguing that black is white is very foolish. Yet those of us who protest at such absurdity often find that we are the ones who are labelled as unreasonable. I have even been accused of "interfering with the language" by the upholders of standards who believe the old chants, and who think that nothing should change from the rules they learnt at school, no matter the degree of misinformation.

Even the distinguished literary critic Alvin Kernan acknowledges that men went a bit too far with their own image building. "Feminists have given us convincing examples of the way that language can control truth," he says, as he comments on the masculinising that the grammarians and dictionary-makers resorted to. The use of *he/man*, he writes, the use of "*man* for people . . . and similar usages have factualised male importance . . . making it seem, because embedded in the language, given and natural that men rather than women are and do everything".[24]

This is the view of the expert who has reviewed and assessed the achievements – and the demise – of print culture.

Male as Norm

The values of print culture have helped to lead to all these discussions. Clearly there are (at least) two sexes and yet, in the interest of forging a single standard, one sex only was made the norm. This is part of the invisibility which has been women's pattern over the last four hundred and fifty years of print. What we need for the future is the full recognition of both women and men – and of all the other groups which make up our community. It's diversity, not uniformity, that we should be looking for and now encoding in our language, in the way that all new technologies have prompted linguistic change.

This is the starting point for Julia Penelope's book, *Speaking Freely*. She too has looked back over the last few hundred years and made the links between language,

22. See Alma Graham, 1975, "The Making of a Non-sexist Dictionary", in Barrie Thorne & Nancy Henley (eds), *Language and Sex: Difference and Dominance*, Newbury House, Mass., p. 62.
23. Casey Miller & Kate Swift, 1976, *Words and Women*, Anchor/Doubleday, New York, pp. 25–6.
24. Kernan, *The Death of Literature*, p. 157.

thought and reality – between the rules men made up, and the way we speak, think, and actualise the world around us. Drilled in the rules that have favoured white, professional men throughout the entire print period, "we find it difficult to think outside the categories and grooves made by men", she states.[25]

Women have not been the only ones whose views and values have been left out. English (and American) have also been the language of the white man.

"Prescriptive grammarians created Standard English, then promoted it as the only 'correct' – that is, valid – version of English," writes Julia Penelope. "The story of English is intertwined with misogyny, elitism, and racism. Prescriptive grammar has served white men's purposes for centuries and its rules have been used to keep enormous social power beyond the reach of the poor and the working classes."[26]

Throughout the print period, standard English came to be used to create a division between the haves and the have-nots. Those who could crack this white, male, professional code were seen as authorities, able to speak correctly, and assumed to be superior.

Those who did not speak properly – who did not have the right accent and table manners – were encouraged to "imitate their betters". On the pretext that they were improving the speech of their pupils, countless teachers forced their charges to learn the standards – or be rated as failures.

For centuries, says Julia Penelope, "teachers have told children they must learn to talk right if they want people to take them seriously. They force us", she says "to abandon dialects, creoles, and other 'sub' or 'non-standard' varieties we speak, substituting Standard English, alleged to be the only acceptable way of spelling and writing."[27]

It is possible to see that there were some good reasons for standardisation, although if the point had been to further communication, then logically speaking, the standard should have been based on the most common use rather than the most privileged. But what we have inherited after more than four hundred years of print is an elaborate system of regulations which have been imposed on the language. The priority for today should be to deregulate.

Dictionary Distortions

One Man's View

Fixing the meaning, spelling and pronunciation of the language is the purpose of a dictionary. It sets down definitions. While no one would deny the invaluable service that dictionaries have provided since Samuel Johnson put the first English one together in 1755, it is clear that dictionaries have also had their drawbacks.

---

25. Julia Penelope, 1990, *Speaking Freely*, Athene, New York, p. xiv.
26. Penelope, *Speaking Freely*, p. xv.
27. Penelope, *Speaking Freely*, p. xviii.

As with spelling, the dictionary soon reveals a gap between its fixed meanings and the changes in the language. As meanings and pronunciations and usages shift (and they always do), the pinned-down version that stands in the dictionary can become further and further removed from the version of the word that the speech community is using. The irony is, it could be that the last place you are likely to find what a word currently means, or how it is pronounced, is in a dictionary.

This isn't the only issue.

In the electronic culture, everything is shifting; meanings are multiple and elusive. There is no longer the same need for standardisation, for definitions of spelling, pronunciation, and meaning. It is no coincidence that word processors have spellcheckers (because we haven't quite got to creative spelling as yet) as well as a thesaurus, which gives alternative words, but generally no dictionary with definitions and set, standardised forms.

Print made the dictionary. The dictionary is the ultimate in standardisation, in the age of standardisation. For Samuel Johnson, the dictionary was a marvellous new tool for refining and understanding the language. For nine years he laboured on the project. In the preface of the first edition, he proudly proclaimed that it was all his own work. It was all a product of his particular experience: a record of his own partiality, or prejudice. This has been yet another drawback of dictionaries.

If it was a white, professional, male view of the world, that you wanted, then Samuel Johnson was your man, and his dictionary was your source of reference. But if you wanted to know whether women had different usages, or whether words had different meanings for the growing audience of mass readers, you wouldn't be able to find out from Samuel Johnson's sample.

You can see why Samuel Johnson thought it was a good idea to make up a dictionary. It would provide some order: it would serve as a bible, a reference point for right and wrong: it would be a force for uniformity. But despite the mammoth task he set himself, it is clear that his dictionary could never have been comprehensive. By its very nature, it was always a record of one man's subjective assessment of meaning.

Samuel Johnson had to start from scratch. Unlike the dictionary-makers who followed him, taking his version as their foundation, Dr Johnson had no models to guide him. He had to begin by listing all the words he knew, and then he had to collect samples of all the ways they had been used. It was all about *his* priorities.

Because he started with his own vocabulary (and that of his friends), the words that were included were those of a particular professional, privileged – and patriarchal – group. Because he made the decision that he would include only those words that were already in print, he further added to the masculinist bias of the entries.

Women – and minority groups – were much less likely to be groomed as writers, or to be published. So you can see why his selection, from the outset, was neither class-neutral nor gender-neutral.

There could have been – would have been – countless usages that Samuel Johnson was not familiar with: expressions and meanings that he had never encountered and therefore didn't include in his records. This doesn't mean that the words he didn't know didn't exist, or that they didn't serve any useful purpose. It simply means that because Samuel Johnson didn't know them, they didn't make it into his dictionary.

Women were not the only ones to be omitted by this practice. Samuel Johnson was not known for his championship of the working classes – or for his support of the upper classes either, for that matter.

There was little love lost between the new Grub Street writers and the aristocracy who were the patrons of culture. By the end of the eighteenth century, the new breed of professional authors had become sufficiently independent and influential to pose a threat to the upper classes who were losing their hold over the world of letters. There was a tussle going on between the aristocracy, who thought they owned the language, and the new writers, who were increasingly staking their claim to the language as their art and their means of earning a livelihood. It was a power struggle.

Not surprisingly, it was a high art and culture that the aristocracy favoured. The classics – often in Latin and Greek rather than translation – were the forms they found uplifting, and they hand-wrote many of their odes and plays and commentaries, and circulated them – "privately" – to one another. They did *not* want to see their names in print, and they despised the idea of earning a living from the world of letters. They saw books, and the mass audience that had developed in response to the printing press, as low culture: and they were disturbed and incensed by the dreadful new scribblers who peddled their stories for profit, and who entertained and excited the common folk.

As one of the new breed of professional scribblers, Dr Johnson was on the side opposing the aristocracy and determined to wrest their power and monopoly from them. This is why you won't find many of the aristocratic usages encoded in his pages. It is also a sign that in the battle over language, Samuel Johnson won. It wasn't the aristocratic version that became the standard for centuries: it was the language of Samuel Johnson and his cronies.

Making the Best of It

To be in charge of meanings is to have a great deal of power in society. (In oral cultures, for example, that's what being an elder, or a wise person amounts to – being the arbiter of community meanings.) Alvin Kernan states the obvious when he says, "Those who can shape language can determine what is reality and what is truth." And, he adds, "A dictionary is one of the primary means of claiming and exercising this power."[28]

Samuel Johnson and his dictionary are no exception. And he started a trend!

The dictionary had started off as a highly biased but *set* list of meanings and pronunciations: it got worse, as one by one, the later dictionary-makers built on the particular and patriarchal code of Samuel Johnson. And because dictionary-makers – like grammarians – have been primarily male, they have continued to add their own masculinist layers.

---

28. Kernan, *The Death of Literature*, p. 157.

This bias persists to the present day, according to internationally acclaimed linguist, Cheris Kramarae. Even where dictionary-makers name the publications that they consulted to make up their lists of words and usages (and many don't include such information), they give the game away. The words of Blacks, women and minority groups are usually not found in the sources that the dictionary-makers turn to.

Sometimes the preface of a dictionary will state that only the *best* authors have been sampled for words and their meanings; sometimes it is only the *most frequently quoted*. But these terms of "best" and "most popular" are not gender-neutral either. Says Cheris Kramarae, "given current cultural practices, not only are men's words more likely to be cited in the mainstream press, but also few dictionary editors seek out print media where women's words would predominate, such as feminist periodicals."[29]

Even that most authoritative source, the *Oxford English Dictionary*, set up a sampling system which left women out. Not that you would know this from the claims that are made for it. For the *OED* was published over a period of forty years (1881–1928) and the 15,487-page dictionary was based on a file of more than a million quotes, from more than five thousand books, which covered a period of seven hundred years.

The trouble is that most of the readers who compiled these examples were assigned a male author to study for meaning and usage. This is why, states Cheris Kramarae bluntly, "men of letters were included in the *OED* – but certainly not the letters of women".[30]

So even when women are in the public sphere – that is, when they are published – their usages are not going to be quoted in any dictionary which samples only male authors. This is how women's history – women's movements and women's achievements – get left out of the record.

Deconstructing the *OED*

The *Shorter Oxford English Dictionary* (1973) for example, has no entry under *rights of woman* and no references to the contributions of Mary Wollstonecraft. This is despite the facts that she coined the term, and that her publications (including *The Rights of Man* as well as *Vindication of The Rights of Woman*) caused a furore in their own time, and have since been labelled the foundation stones of the modern women's movement and its quest for human rights for all.

Just to show how biased the entries are, I should point out that in the 1973 *Shorter OED*, under *rights* there is a quote from Shakespeare; there is also one from Thomas Otway – whoever he may be – which is "Your Sex Was never in the right, you're always false, or silly." This entry could serve as the subtitle for the entire sexist genre of dictionaries throughout print culture.

In the 1973 edition of the *OED*, only 70 of 1525 people quoted are female. That's less than 5 per cent. And that is not the full picture. The 1455 men who are quoted

---

29. See Cheris Kramarae *et al.*, 1985, *A Feminist Dictionary*, Pandora, London, p. 2.
30. Cheris Kramarae, 1992, "Punctuating the Dictionary", *International Journal of Sociology of Language*, 94, p. 142.

include those like Dickens and Pepys and Shakespeare who are cited again and again; while the 70 women are quoted very infrequently. So the ratio of female to male meanings could still be as high as one to ninety-nine. This does not represent the ratio between women's words and meanings, and those of men!

Sometimes it is difficult to accept that prestigious male dictionary-makers could so blindly cling to the male standard. Philip Gove (*Websters Third International Dictionary*) boasts that his dictionary is comprehensive, that it cost an enormous amount, and that it was put together by highly qualified staff who consulted thousands of print sources. But when he adds that these consisted of other dictionaries and the Bible – where women have been systematically excluded – his claim becomes something of a joke. In believing that almost all-male sources can be comprehensive, this editor also shows that not much has changed since the prescriptive grammarians put forward their argument that *man* stands for *woman* on the grounds that the male is more comprehensive.

As a female, it is pretty infuriating to be confronted with standards that have come from men checking with other men, and recording their meanings as definitive, on experiences that are totally outside their male experience. They make no acknowledgement that there's anything wrong with their method. Rather, the myth has been quite the reverse: that the dictionary is objective, authoritative, impartial, definitive and correct; that women have got it wrong if they think the dictionary contains any errors. And this is not the end of the faults and the frustrations.

In language, the bias and silencing of women functions at the most basic and pervasive levels. There are many more words for men for example, and more of them are positive. If anyone wants to know why this is so, they have only to look at the dictionary-makers to get some idea of the priorities.

Men have managed to make themselves look better, even when it borders on the absurd. So, for example, Julia Penelope found that there are approximately three hundred words for *prostitute* in the establishment sources, but only a very few for *whoremonger.* (It's not actually a word that you would use in everyday conversation, whereas *prostitute* is a common term.)

Now this preponderance of words for what women do cannot be a realistic assessment. If there really were three hundred prostitutes for every whoremonger, there would be a lot of sex workers unable to make a living. It would be much more realistic if the number of dictionary entries were reversed, and there were three hundred words for whoremonger, and only a few for prostitute. No doubt if women had been consulted they could have come up with many more terms for the role played by men. The results wouldn't necessarily make men look good.

The words that have been chosen for inclusion, and the meanings that have been selected for them, give the game away when it comes to the identity – and the self-promotion – of the dictionary-makers. The men who were in charge took themselves as the starting point, the norm, the benchmark: they measured the rest of the world against their own standard. Not surprisingly, they soon defined women as non-men, as non-standard, as negative, deviant and "other". This arbitrary but all-pervasive *man-made* inequality is recorded in the average dictionary for all to see. And use.

Women shouldn't be distressed by the demise of the dictionary. Any ideas about "a drop in standards" have appeal for those who have been penalised by the operation of a single, male standard.

Dictionaries have never been friends of women or minorities, as Cheris Kramarae has pointed out, and women would be better off without the definitions which constantly brand them as deficient. If we could get rid of the so-called "correct" forms which have invisibilised women throughout the print period, the opportunities would be there for women to define themselves, to provide their own meanings for their own experiences. They could decide their own truths and realities, as the experts – like Alvin Kernan – occasionally admit that men have been able to do for centuries.

Women's Words

Already steps have been taken in this direction; there is *A Feminist Dictionary* (Cheris Kramarae, Paula Treichler and Ann Russo, 1985) as well as *A Women's Thesaurus* (Mary Ellen Capek, 1987) and *Websters First New Intergalactic Wickedary of the English Language conjured by Mary Daly in cahoots with Jane Caputi* along with *Womanwords* (Jane Mills, 1989) and *Womansword* (Kittredge Cherry, 1989). In these volumes you will find usages and meanings that are positive about women and critical of men – in contrast to the traditional publications. The men's dictionary has been one of the best-kept records of misogyny.

If there is one thing that they have been able to do while they have been in charge of the standards, it is to give themselves a good press: wife-beater doesn't even rate a mention in the 1973 *OED*, though Frances Power Cobbe was one of the many to use the term in 1868.[31]

The standard representation of women in dictionaries has been the negative stereotype. There are literally thousands of examples that could be quoted, but a typical one (from *Random House Dictionary*) is an illustration for the word *nerves*. "Women with shrill voices get on his nerves" states the dictionary example.

It is not just that women are consistently *portrayed* as shrill, as sex objects, and as stupid; Cheris Karmarae points out that, because women's meanings are not quoted within dictionaries, women themselves are not seen as makers of culture, as active language users. Rather, another stereotype of women is conveyed; as that of "mere receivers and transmitters of the code and hence incapable of making original contributions to the language".[32] Nowhere is this more obvious than in the case of the new words that women make.

There's only one way to get words into the language: to make them up. It's happening all the time: that's the way language stays alive and dynamic. When men do it,

---

31. See Dale Spender, 1994, *Weddings and Wives*, Penguin, Ringwood, Vic., for further elaboration.
32. Kramarae *et al.*, *A Feminist Dictionary*, p. 3.

their coinages are taken seriously and soon find acceptance at every level from slang to scholarship. But when, during the print era, women have made up words that have specific meanings for them – words such as *herstory* or *malestream* for example – they are generally treated as a joke. Such words are regarded as a sign that there's something wrong with the women who are trying to interfere with – and debase – the language.

There isn't even a limbo state where women's clever coinages can hang out until they acquire some acceptability and gain entry. By definition, slang is the province of men.

Yet *herstory* and *malestream* (coined by Mary O'Brien) are very meaningful terms for women. They name things which happen to women every day as they are confronted with male versions of events, and find themselves dismissed for being outside the male-mainstream. Along with words like *malefunction* and *Manglish* – which are very useful – there are terms which no woman should be without: *testeria* and *penisolence* (Julia Penelope) are indispensable (and require no elaboration!).

Words make a difference. In her article on menopause, Brinlee Kramer[32] makes it clear that one of the reasons that the state is not taken to be "real" is because it's outside men's experience and they have not needed to make up the words which label the phenomenon. She makes good the deficit with such coinages as:

> *dysmedicocognition:* doctors' ignorance thereof
> *pyrosomnia:* interruptions to sleep caused by hot flushes
> *languor climacteria:* physical exhaustion associated with menopause
> *polycarbophagia:* from "poly" (lots) and "phagia" (eating), it sounds infinitely better than
> "pigging out"

These wonderful and empowering words fulfil the requirement of being in print; they all occur (more than once) in books that are on my shelves. But most of them are books that your average male dictionary-maker is not going to look up.

Occasionally, it happens that a woman's coinage becomes so much part of the culture that it cannot be ignored. But it is interesting to see how it can lose its female perspective when it is entered into a traditional dictionary.

Take the word *sexism* for example, and the definition it has been given in *The Macquarie Dictionary* (1986). It started out as a pretty radical term. I can even remember the time when it wasn't for use in polite circles. *A Feminist Dictionary* defines it as "A social relationship in which males have authority over females", which is still a fairly strong label. But in the mainstream dictionary, it has become "the upholding of sexist attitudes" which sounds almost chivalrous. And *sexist attitudes* themselves have also lost much of their force: in *The Macquarie Dictionary*, the term is held to mean

---

32. Brinlee Kramer, 1993, "The Language of Menopause: Running out of Eggs", paper presented at Fifth International Interdisciplinary Conference on Women, San José, Costa Rica, February.

"to stereotype a person according to gender and sexual preference". It's certainly not how I would define sexism; or sexist attitudes. Or even sexists themselves.

Clearly these sexist definitions do not fit with the expressions of an average female dictionary user.

"Dear Dale Spender" begins a letter I received in December 1993:

> There is something I have been concerned about for several months and I decided to write to you about it.
>
> I purchased a *Macquarie Dictionary* and *Thesaurus* earlier in the year and as I am interested in words, spellings etc., I thought I'd be very happy owning it. However, I became quite shocked when looking up . . . these entries:
> *bitch:* female dog, woman
> *gossip:* given to idle chatter, especially a woman
> This started me looking everywhere to find some derogatory terms bestowed on men but I could find nothing. Another thing that annoyed me was
> *weak:* pertaining to female, and then vice versa,
> *female:* weak,
> which I think is a put-down, and a myth when you look around at women's capabilities.
>
> What concerns me is that this dictionary . . . is I believe being used in schools and as I have five grand-daughters I am sickened to think that this thing will be handed down the generations of young boys influencing their attitudes to women, which has gone on for centuries. I wrote to the Macquarie publishers about the *female–weak* subject and just received a nice reply but saying that they wouldn't and couldn't do anything about it.
>
> Why on this earth can't something be done about it? Surely it's time definite steps were taken to reduce the humiliation girls and women have to suffer from day to day in our society. I was so angry I have actually thrown the book away and I worry continually about it being used in schools. No wonder men find it so easy to trivialise and degrade women when it was amongst their school books as boys.
>
> Can you do anything about it? Or can you point it out to the Education Department and have something done quite soon or forward my letter to them showing my concern for future generations of young women?
>
> > Yours sincerely,
> > Ms Dawn L. Painter

Well, I can tell Ms Painter that dictionaries won't matter nearly as much in future. She is quite right – throughout the print period they have not served women at all well, and I for one certainly won't be lamenting their departure. I'm quite happy to settle for a spellchecker and a thesaurus (though, with the exception of *A Women's Thesaurus*, they too could do with more female-positive alternatives).

I won't find it a hardship to do without the record of men's view of the world, and women's place within it, which is what the standard dictionary amounts to.

2

THE CLAIMS OF LITERATURE

Exploding the Canon

A Good Story

The standardisation that went with print wasn't limited to typeface or spelling or grammar: literature has also had its share of rules and regulations, despite its reputation for creativity and imagination. The idea of a literary canon – a list of what is in and what is out of the great tradition – points to the way literature has been divided into the good and the bad, the norm and the deficient. ("Canon" by the way, is defined by the *Shorter Oxford English Dictionary* as "a standard of judgement" and "the list of books . . . accepted as genuine and inspired".)

Of course, when it came to literature, the people who were making the decisions about what was worthy and what was not were none other than those same white, professional males. It should come as no surprise to find that the canon they chose, the standard they set, was that of white professional men. When in 1952 a series called *The Great Books of the Western World* was published, it did not include a woman or a Black writer. Fifty-four volumes, seventy-four authors, 443 works, and not a woman or Black could meet the standard! The judges insisted that this had nothing to do with race or sex – it just "happened" that all the best books in the world had been written by white men. Their arguments did not appear intellectually rigorous, and the supposed coincidence was not at all convincing.

What we know as "literature" is a product of print. There was no great creative literature before the advent of the printing press – unless, of course, you want to put the Bible in this imaginative category. There probably will not be any such entity as "literature" after the presses give way to the Internet as the major source of publication.

In the pre-print era, along with the biblical texts, there were sermons, religious commentaries and so on, but these were not seen as literature as we have come to know it. They were not regarded as original contributions of the human mind. Far from it. They were held up as evidence of divine inspiration, as truths which came from God and were *not* contaminated by human invention.

So the idea of literature as fiction – human stories made up and expressed in poetry, drama, fantasies, etc. – didn't make its appearance until the age of standardisation. And what we know and understand about this particular concept of literature is likely to change as standardisation gives way to pluralism, and as print makes room for image.

Literature as Criticism

Despite all the claims for literature as the product of the human spirit and the highest form of representation, its beginnings were much more humble. Throughout the eighteenth and nineteenth centuries, when scientific rationalism and the factory system became the order of the day, there was a pocket of opposition to this utilitarianism. The romantics began to insist that there was more to life than making profits and that there was not always a rational explanation for everything that happened.

So the Romantic poets began to look for beauty on Grecian urns and among the daffodils, while the early Gothic novelists – Ann Radcliffe the most renowned among them[1] – took to writing stories in which the obvious or scientifically rational deduction was definitely not the solution.

In retrospect, it's ironic that this concept of literature as inspirational and life-enhancing, began as a criticism of the money and machines of the Industrial Revolution. Because the art of these new creative writers was entirely dependent upon the forces they despised; without the profit motive and the mechanised capacity of the printing press, there wouldn't have been any literature. This has led the eminent literary critic, Alvin Kernan, to accuse the profession of "biting the hand that feeds it" – both past and present.[2] He goes on to charge contemporary literary academics with bringing about the downfall of the discipline with their destructive, deconstructive tactics.

But literature has never really had a firm foothold in the university community. Apart from the fact that it is a relatively recent arrival, it has generally been plagued by doubts about what the subject is, and what claims can be made for it.

For the Romantics, however, the very idea that a body of creative writing – with the exalted status of literature – could be taught within the university would have been beyond their wildest dreams. In their day, literature was an alternative, in the nature of a protest movement, rather than an "ideological arm of the establishment" (a charge levelled at it by some critics today). Because it was classified as lacking in substance, scholarship and scientific rigour, the task of having English literature become part of the university curriculum posed a problem. It was not until well into the twentieth century that the University of Cambridge, for example, admitted English to the hallowed halls of learning. And this was only after much lobbying, and a lot of wheeling and dealing.

---

1. Ann Radcliffe, 1764–1823; *The Mysteries of Udolpho* (1794) and *The Italian* (1797) are among the best known and most highly acclaimed of her works. For further information see Dale Spender, 1986, *Mothers of the Novel: 100 Good Women Writers Before Jane Austen*, Pandora, London.
2. Alvin Kernan, 1990, *The Death of Literature*, Yale University Press, p. 25.

University Entry

The first attempt to get English literature introduced into Oxford University was in 1887, and the move was greeted with resistance and ridicule. The case against it, put forward by the university establishment, was straightforward and simple: not much more than one hundred years ago, English was not up to standard.

Today, we have come to accept English as the leading world language, the international language for which all sorts of claims for superiority are made. It comes as a shock to find that only a century ago English itself was considered too crude and unrefined to warrant study. The elite and pretentious (or, as some would say, the ignorant and self-important) scholars of the time acknowledged that they could see why students needed to learn Latin and Greek and to study the illuminating classics, but what was there to learn about English? Everyone could already speak it. After all, it was the vulgar tongue which hadn't even produced a first-class literature.

It is interesting to note the parallels here with the introduction of Australian literature (or any other colonial literatures seeking university status). Just as the nineteenth-century champions of Latin and Greek were dismissive of a native English tradition, so the later champions of Great English Literature were equally dismissive of a native Australian tradition.[3]

Of course, this is one of the by-products of setting up a standard: everything that is not included immediately becomes *non*-standard and is soon regarded as deficient, deviant, not the real thing. And once English literature replaced the classics as *the* standard, it became the yardstick against which colonial literatures were judged as parochial, primitive and plainly inferior.

Needless to say, this had nothing to do with the literatures. Like many of the standards associated with print, this one says more about the mindset that divides difference into good and bad, than it does about the contributions of the many literatures.

English literature didn't begin to set itself up as the standard until 1894, when it was introduced at Oxford University. (Its swift and spectacular success has to be acknowledged.) Because there were so many initial objections to the lightweight nature of the subject, much of the emphasis from the outset was on dressing up English literature so that it looked more substantial. Then it could be held to be "true" and could command authority.

Have you ever wondered why the student versions of Shakespeare and other English "greats" had footnotes, glossaries and all that information about setting, plot, etc. at the front of the book? Because that's how Latin and Greek texts were set out, and the English lobby wanted it to look as though their literature was just as substantial, and generated just as much scholarship, as the prestigious classics.

---

3. The Australian writers Florence James and Dymphna Cusack, who attended Sydney University in the early 1920s, were incensed that Australian history and literature were regarded as unworthy of study. With Miles Franklin, they protested over the years – often in their writing – against this cultural imperialism. Even in the 1960s and 1970s (and 1980s and 1990s) there were still academics who believed that Australian literature was not up to standard – the "cultural cringe" curriculumites.

When the moves were made to get English into the curriculum, the prevailing orthodoxy in the universities was scientific thinking. The only information that was considered real and reliable was that which met the requirements of scientific dogma. So the English literature lobby faced the challenge of making their subject seem more like the mechanics of Newton's laws, rather than odes, elegies, and products of human imagination.

There was no point in the defenders of English arguing that works in the native language were "the poor man's classics" and that they would expand the soul and civilise the masses. The establishment was not impressed. If it wasn't a body of rules that could be tested, the in-house scholars declared, then it wasn't academically respectable, and had no place in a university.

Professor Augustus Freeman was not to be swayed by all the persuasive, literary-romantic arguments that the reading of poetry would make for better individuals and a more humane world. He said it all in an article in the London *Times*: "There are many things fit for a man's personal study, which are not fit for University examinations," he explained to the English literature advocates. Literature was one of the things that didn't qualify. "We are told that literature 'cultivates the taste, educates the sympathies, enlarges the mind.' Excellent results, against which no one has a word to say. Only we cannot examine in taste and sympathies."[4]

In other words, if you couldn't measure it, if you couldn't have laws or regulations which could be standardised, memorised, and examined like theorems, then the subject couldn't qualify as university material.

So it seemed that there was no room for English.

Not to be defeated, however, the supporters of English determined to satisfy the university criteria. They would *make* English into a body of rules and rituals that would do any academic board justice.

Rules of the Game

Old English as English

It was all achieved very simply. A body of rules about English that had already been formulated was put forward for vetting. All the information about the mechanics of the language – from grammar to vowel movements – was paraded, and called English. Which is why from its earliest days the university study of literature was "a branch of philology, based firmly on the study of dead languages which surely needed to be taught and certainly could be examined."[5] Equally surely, it had nothing to do with the vision of literature – the ideas, intentions, meanings and styles – that the Romantics had in mind.

---

4. London *Times*, 8 June 1887.
5. Kernan, *The Death of Literature*, p. 39.

So Anglo-Saxon, Old Norse and Old English found themselves well received in university circles. Not because anyone was necessarily interested in them, or profited from them, or because they made any worthwhile contribution to the sum total of knowledge. But because they could be organised, standardised, memorised, regurgitated – tested. Good educational material.[6]

This form of English did not, of course, satisfy the champions of the great literary tradition. They were soon claiming – with some justification – that the study of literature was "little more than hunting sound changes through primitive Germanic forests"[7] – a sentiment with which generations of English students could readily identify.

For decades there were no great changes in this pretence about English. When I was a university student thirty years ago, the English Honours course was based upon Anglo-Saxon. Study of this dead language was ruled to be rigorous and "ranking"; it was supposed to sort the students into the successes and failures. There could be no career in English for those who didn't learn the vowel shifts of past centuries.

That ruled me out. I wanted to study books (as do some equally frustrated students today). But my way was barred by the boring and banal study of the readily examinable rules of a language that had long passed its use-by date. It was more than frustrating to come up against a brick wall when I argued that Grimm's Law (and the first sound shift) had nothing to do with my desire to analyse the irony in the novels of Jane Austen.

Of course, bit by bit, English literature did make its way into the curriculum, but in retrospect, it was never in a very strong position. It was always trying to meet the demands of the standardisation pressure group. This is one of the reasons that the canon materialised. It contained the tried and true masters, who were to be the models; their work was held up for emulation.

Many of the certainties associated with scientific rationalism soon found their way across to English, with some very silly consequences. In the case of Old English – where the language had become fixed and the rules and regulations were not inconveniently challenged by a community of speakers using the language – the difficulties were probably not so great. But in trying to apply the standards of physics to literature, some fantastically foolish moves were made.

The Test of Greatness

The force with which literary rulings were pronounced left the profession open to mockery and ridicule. It was as if a litmus test had been devised to determine who was outside and who was in the canon. Woe betide any unsuspecting student who got it wrong and (in the absence of any women worthies) wrote an essay in praise of John Donne, for example, *after* it had been decreed that he no longer met the standard for literary greatness.

---

6. See Chapter 5 for further critiques of this type of knowledge.
7. Kernan, *The Death of Literature*, p. 39.

On the one hand, there was a conviction in the profession that it could tell who made the grade as a writer and who did not, and that this determination was absolute. But on the other hand, even the most unsophisticated literary historian could see that one generation's heroes were the next one's hates. So one generation would hold up the Romantic poets as ideal authors, only for the next to condemn them as untalented, wet and wimpish. Then in turn the Romantics would be replaced by the true rational genius of the Augustans. Etc., etc. So much for truth, and quality – and absolute standards.

There is no doubt that there has been a lot of puff and pretension in the pronouncements of the literary establishment. For some people the opportunity to show that the Emperor has no clothes has proved tempting. Australia has its own famous exposé incident in the annals of English literature: the Ern Malley affair.

In the course of one afternoon, in a military barracks in Melbourne, Corporal Harold Stewart and Lieutenant James McAuley "constructed" the poetic works of one Ern Malley, a genius who had met an early death. The poems, along with a covering letter from Ern's sister Ethel, were despatched to Max Harris, the bright young editor of the literary magazine *Angry Penguins*.

To cut a long story short,[8] these "poems" were received with great enthusiasm by an entire artistic circle. They were published to great acclaim. When the real authors did stand up, it was clear that "The Ern Malley Affair was without question, the literary hoax of the century."[9]

Such scandals about the standards of the literary elite have done little to promote a positive image of the profession, or to provide it with a defence at the present time when the entire literary enterprise is under threat.

But such scandals do help to explain why literary critics have tried to take refuge in rules and mechanics such as those of dead languages (or literary theory). To deal in fixed bodies of knowledge is to be much less vulnerable to charges of falseness or foolishness.

It was hard to be proven wrong in the framework of Old Norse. And while there are differences, it is also hard to be proven wrong in the framework of the current "new criticism". A body of knowledge that has as its starting point that authors can't be trusted to know what they mean (so there's no fear of contradiction from that quarter) gives itself a lot of leeway. And any discipline that holds that there are as many interpretations of a text as there are readers is not going to founder on the rock of truth and certainty. By speaking in terms of semiotics, few of today's literary critics are going to find themselves in the hot seat defending standards of greatness and universality. Though they might find themselves facing other problems.

---

8. For an excellent coverage of the episode, see Michael Heyward, 1993, *The Ern Malley Affair*, University of Queensland Press, St Lucia, Qld.
9. See Robert Hughes, "Introduction", in Heyward, *The Ern Malley Affair*, p. xvii.

Downfall and Dirty Tricks

Professional Decline

All this is by way of introduction to the fact that, only a century after they got a foot in the door, English literature departments the world over are in decline. Serious literature now has an audience limited almost entirely to the professionals, and even their ranks are diminishing. Figures for the United States show that, while the number of bachelor's degrees given in the last decades has increased by 88 per cent, degrees in English are down by 22 per cent.[10]

Alvin Kernan blames many of the literary academics themselves for bringing about their own demise. But while you might argue that the professionals could have done a better job in defending their empire, the real force which has undermined English literature is the new media. English literature is now culturally superseded.

People now watch television, listen to the radio, or dial up a database on their modem, rather than read books. The very form of the book is becoming passé in a world where

> television is transforming everything it touches – politics, news, religion – where increasing numbers of citizens have great difficulty in reading even simple texts, where creativity and plagiarism are increasingly hard to define, where advertising and image-making have captured the language.[11]

There is a literary crisis because print, the medium on which literature is based, is waning in importance and influence as electronic communications become more prevalent and popular. But Alvin Kernan is by no means alone when he asserts that, instead of rising to the challenge and finding new and vital ways of redefining literature in an image-based world, too many literary academics have hastened the process of cultural obsolescence.

Going through the addresses of the presidents of the Modern Languages Association in the United States, Gerald Graff concludes that literary scholars have never had a coherent or consistent view of what it is that they are doing, and that too much energy has gone – and still goes – into professional sniping and sterile self-justification. According to Gerald Graff, the history of literature in the university has been "a lamentable failure made up of endless confusion and argument, to the great detriment of the profession and the diminishment of the subject".[12]

Alvin Kernan agrees: he summarises the history of cliques and cabals, controversies and crazes that have characterised the presence of English in the universities.

---

10. See Kernan, *The Death of Literature*, p. 84; this decline in English as literature is, of course, in contrast to the increase in the teaching of English as a second language.
11. Kernan, *The Death of Literature*, p. 10.
12. Kernan, *The Death of Literature*, p. 42.

For a time, literary texts were read, as decreed at Oxford, as evidence of the history of the language, then as documents in social history, then as chapters in the biography of the writers of the "life and times" variety. At later times, in the American new criticism and the *Scrutiny* movement led in England by F. R. Leavis, literature rejected its dependence on linguistics and history, and declared its autonomy as a moral and aesthetic reality existing in its own right and providing its own special kinds of truth and understanding. Formalism has now passed in turn, and at the moment the most energetic types of literary activity – feminist, deconstructive, new historicist, Marxist and psychoanalytic – share a social conception of literature described succinctly by Terry Eagleton as "modes of feeling, valuing, perceiving and believing which has some kind of relation to the maintenance and reproduction of social power."[13]

The latest development – that of literary theory – is the last one to bring English into disrepute in Alvin Kernan's book. It's possible that many would agree with him. I am sure that my university class was not the only one to have been on the receiving end of some of the petty rivalries, schisms – and towering certainties – of the profession's activities. Well do I remember my final year as an English student, when the new professor arrived; he had completely different views about the subject, and informed the class that it was too late for us. We were a lost cause. As we would never be able to learn the basic principles in the short time that was left, there was no point in him teaching us James Joyce. (For small mercies, much thanks!)

But this was hardly good news for the graduates, or comforting to those of us who went out as high school English teachers. To be treated so cavalierly and with such contempt did little to enlist our loyalty or inspire confidence in the profession. When today a few leading literary academics declare that some of their colleagues have contributed mightily to doing themselves out of a job, classes upon classes of student-victims could be found to concur.

Certainly, today's students are voting with their feet. The number who want to do English (a subject which now almost always encompasses some of the mechanics – semiotics, deconstruction, literary theory) is decreasing every year, so that the profession is shrinking dramatically.[14]

But it is not the student enrolment alone that Alvin Kernan takes as a sign of the malaise; it is the fact that English literature has little presence or relevance these days outside university walls: "almost no one outside the university reads the art novel and the poetry of our time, or believes that they have any bearing on the serious business of the world".[15]

Kernan maintains that in the face of the threat of obsolescence, the English literature lobby could have done much better. It could have put up more of a fight to defend its territory. It is "a time of trial", he argues, and yet "literature has only feebly justified

---

13. Kernan, *The Death of Literature*, p. 42.
14. There is some evidence that the reverse is true in Women's Studies – perhaps because only recently have women been able to study women's writing, there is much enthusiasm for such courses.
15. Kernan, *The Death of Literature*, p. 33.

itself". Instead of going out and converting the population to the greater need for books, it has behaved in a foolhardy and self-destructive manner. "Literary people tend to think, despite the evidence, that literature is of crucial importance to society and take its presence in the university curriculum for granted," he states bitterly.[16] Many in the profession see themselves as upholding the standards of civilisation and humanity, and it is beyond their comprehension that their expertise might no longer be necessary.

To Alvin Kernan, the results of such a policy are obvious. Literature no longer has a dynamic identity; "in most places university departments have decayed into service departments providing support work in reading and writing for the least qualified students."[17] This is the death of literature as far as he is concerned. And he holds many of the new critics responsible for the demise.

Taking the "Con" out of Deconstruction[18]

Alvin Kernan sees the contemporary literary theorists as having turned treacherously against their own subject. He accuses them of contributing to the downfall of literature by discrediting the great tradition. "The great books which . . . hitherto formed the basis of liberal education [have been] denounced as elitist, Eurocentric, the tools of imperialism," he says: and he goes on to show what happens when practitioners turn viciously upon their own discipline.

"Under this kind of pressure, the faculty and administration [have] agreed to replace such writers as Homer and Dickens with books like Simone de Beauvoir's *Second Sex*."[19] This is obviously a terrible thing. To Alvin Kernan – and other academics who deplore the dropping of the standard – it is those who censure the racist, sexist and imperialist bias of the canon who have to answer for the present unfortunate state of the discipline.

But to condemn the critics who have branded the canon racist and sexist for the death of literature is to blame the messenger rather than the message.

The canon is indisputably racist and sexist. It is elitist, imperialist, and Western in its orientation. It is another collection of white, male, professional expressions that have been officially stamped as the norm. Since the first days at the University of Oxford, a circle of educated men have paid attention to the writings of other men, and they have elevated those that they like into the standard for appreciation. And they continue to insist on these works as the reference point for all else that is written.

There can be no better example of their prejudices at work than in *The Great Books of the Western World*. This selection is the substance of the canon. For the first time, the 1990 edition included women – four of them. But in the sixty-volume series, no Black authors were represented. This led Henry Louis Gates to state "there's still

---

16. Kernan, *The Death of Literature*, p. 10.
17. Kernan, *The Death of Literature*, p. 59.
18. With grateful thanks to Margaret Atwood, 1994, "Not Just a Pretty Face", *Women's Review of Books*, 11, 4, January, pp. 6–7.
19. Kernan, *The Death of Literature*, pp. 3–4.

a 'whites only' sign on what precisely constitutes a great thinker".[20] There is still a "male only" sign on what the men think constitutes a great writer!

Clearly, some of the old guard have reached the stage where they believe in the myths of their own making. They have come to think that there is something absolute and objective about *their* literary standards. They think that any modifications – the inclusion of any other voices such as those of women or Blacks – will result in sinful relativism. One thing will be as good as another: Blacks will be as good as whites, women as good as men, and critics as good as authors. The notion of excellence – of those who are out and those who are in – will cease to be meaningful.

At least, this is the way that they phrase their arguments for the preservation of the white male literary code. To those who are part of it, it doesn't feel as though the great tradition is the best of what like-minded individuals have written and what they happen to have read. It feels like the best there is. But this is, of course, a very limited and skewed view of the world, and of literature. It also represents a very low standard of scholarship.

In *The Writing or the Sex? Or Why You Don't Have to Read Women's Writing to Know It Is No Good*,[21] it is readily demonstrated that many of the professionals who dismiss women and Black authors do not even follow the basic rules of logic. It isn't that they have read all the women writers and found them not up to scratch. It is that they believe that women are not up to standard, so that only the very exceptional among them (four to be precise, if the 1990 *Great Books* series is to be trusted) could produce literature worthy of acceptance and appreciation. They don't have to read the work of women and Blacks to know that it doesn't qualify for inclusion.

It is this false reasoning and intellectual dishonesty which is more likely to bring English literature into disrepute, than any attempts to name its flaws, or to put its house in order.

In *The Death of Literature*, Alvin Kernan states that his "argument is, to put it simply, that we are watching the complex transformations of a social institution [literature] in a time of radical, political, technological and social change".[22] But he still sounds as though he thinks that the new critics (particularly the feminists and the deconstructionists) should take more of a share of the blame. One of his greatest grievances is that with all their new-found influence, the critics are disturbing the "natural order".

It used to be the critics who were the humble servants of literature, but these days there has been a complete reversal of the literary values.[23] We now have critics who think that their contributions are as good as – if not better than – those of the imaginative authors.

---

20. *New York Times*, 25 October 1990.
21. See Dale Spender, 1989, *The Writing or the Sex? Or Why You Don't Have to Read Women's Writing to Know It Is No Good*, Athene, New York.
22. Kernan, *The Death of Literature*, p. 10.
23. Kernan, *The Death of Literature*, p. 8.

Where once critics and reviewers wrote anonymously – the practice until very recently – we now have critics not only signing their names but becoming identities, even celebrities, in their own right. And while Alvin Kernan might have given some grudging respect to the Sydney-based clique of Don Anderson & Co.,[24] it doesn't take much imagination or perspicacity to know what he would think of Antonella Gambotto, who has taken the art of being a reviewer and critic into the realm of the cult figure.

> This journalist cum putative novelist cum occasional TV interviewer cum frequent scourge of the rather rich and modestly celebrated is poised, she hopes, to become a hot number on our literary scene . . . and the wailing from victims of her acerbic prose is but background babble as she pursues her ambitions.

So writes Murray Waldren in the *Australian*.[25] Antonella Gambotto is famous for being a flamboyant and ferocious critic with "a merciless eye for minutiae and a matador's instinct for the kill"; and for six years she has been working on a novel. Alvin Kernan, who puts literature first, finds this a reversal of the natural order; the novel from the critic is a bit like the book from the film.

In an electronic and image-based age, of course, those with the image (and there is no doubt that Antonella Gambotto has got it) are more likely to be feted than those who make their mark with print. And this is a measure of the fall of literature.

In relation to the academic critics who are putting themselves above the creative writer, Alvin Kernan laments the way they have "seized" control, proclaimed the truth of theory, and begun to treat the great books as nothing other than a database to illustrate their own aggressive platforms of feminism, Marxism or deconstructionism.

Ending a Great Tradition

What Alvin Kernan doesn't say is that the standard he is defending is the aggressive platform of white, male, Western supremacy. Women, Blacks, and writers of post-colonial literatures, for example, simply want the white men to move over; they want a Great Tradition that includes a diversity of voices, and resonates a range of experiences, from around the world. Not just those of white men in praise of one another.

But, of course, if everyone was let in, there couldn't be a canon. Not as we have known if for the past few decades. The whole point of it is to be exclusive, to exalt excellence. Thus the hoi polloi are kept out of the running, and there are standards to teach to the next generation.

No doubt there is much that could be criticised about the new critics, but we should not let the old champions of literature off the hook. Their continued commitment to the single, white, male standard is increasingly indefensible. It brings them and their cause into disrepute.

---

24. Don Anderson teaches at a Sydney university and writes regularly for the major newspaper in that city.
25. Murray Waldren, 1994, "In the Blood", *Australian Magazine*, March, p. 19.

When asked about the absence of Blacks and women other than the famous four (Jane Austen, Willa Cather, George Eliot and Virginia Woolf), the editor of the 1990 edition of *The Great Books* was adamant. "They came very late in the picture," he said dismissively. And then, in one of the most bare-faced lies in literary history, he went on to assert that there were almost no women writers before the nineteenth century.

Now I and countless others know that in English, anyway, there were more women novelists published in the eighteenth century than there were men. Many courses have been set up on this topic, in many universities, and many books have been written on this subject. There are basic reference works such as Janet Todd's *Dictionary of British and American Women Writers, 1660–1800*, not to mention the internationally acclaimed contributions of Joanna Russ, Jane Spencer, Mary Anne Schofield and Cecelia Macheski,[26] to name but a few. So firm was women's hold over the novel during this period that it was men who were likely to use female pseudonyms in order to get published.[27]

If the editor of *The Great Books* (along with his panel of twenty-three top advisers – which includes one woman, Leonie Kramer) doesn't know this, he shouldn't be in the business.

And if he does?

The printing press broke the monks' monopoly on information: it democratised reading and allowed the masses a measure of autonomy to find out for themselves, to draw their own conclusions. And now the electronic networks are breaking the hold that professionals have had over literature. They are democratising authorship. They are allowing everyone to have a voice, to speak for themselves – literary theorists among them.

As the monks deplored the end of their religious truths, so too can many of the literary profession deplore the end of their truth, their brand of excellence, their white, male standard.

Literary theory is itself a response to the new technologies, an attempt to account – albeit pretentiously or even pitiably – for the breakdown of the boundary between writer and reader. We are moving from mass readership to mass authorship. In the words of George Landow and Paul Delany, two of the critics who bridge the area between the old literature and the new hypertext, the concepts of literary theory are virtually embodied in the new multimedia.[28]

---

26. Janet Todd, (ed.) 1987, *A Dictionary of British and American Women Writers 1660–1800*, Methuen, London; Joanna Russ, 1984, *How to Suppress Women's Writing*, The Women's Press, London; Jane Spencer, 1986, *The Rise of the Woman Novelist: From Aphra Behn to Jane Austen*, Basil Blackwood, Oxford; Mary Anne Schofield & Cecelia Macheski (eds), 1986, *Fetter'd or Free? British Women Novelists 1670–1815*, Ohio University Press, Athens, Ohio; and Dale Spender (ed.), 1992, *Living by the Pen: Early British Women Writers*, Teachers College Press, New York.

27. See Spender, *Mothers of the Novel*, for further discussion.

28. George P. Landow & Paul Delany, "Hypertext, Hypermedia and Literary Studies: The State of the Art", in Landow & Delany (eds), 1991, *Hypermedia and Literary Studies*, MIT Press, p. 6; it is a mark of the rate of change that I have substituted the term multimedia for hypertext, which is now relatively obsolete.

The literary canon came with print. It was an attempt to establish a benchmark, to set up a standard, and as with grammar and the dictionary, its rules were those of white, professional men. Because there is no place for a single, exclusive standard in the new global networks, the canon and much of the justification for literature now have little credibility.

It remains to be seen what happens to the white professional men.

3

READERS

The Professional Reader

Reading the Lesson

Before the printing press and the book made their appearance, only a very few people were readers: the priests and the princes.

Nearly all the manuscripts that they read were in Latin, so it was pretty easy to ensure that "the word" could be kept in the right hands. Even if they did manage to learn to read, the masses wouldn't be able to make sense of the Latin code.

Not that it was likely that anyone outside the elite would become literate. Reading was considered a difficult skill to acquire; you had to have special talent and then be trained for a long period of time if you wanted to master the habit.

This is in contrast to what we think about reading now: in modern times we expect everyone to be able to read. We have Education Acts which require every member of society to be literate. We have made learning to read a universal human right.

Before the advent of print, reading was a profession, like law or medicine today. You had to go to a special place to study reading, to learn how to do it properly, to pass the exams and become qualified. Only the best and brightest could hope to be so talented. Unless you were a member of a noble family with your own private tutor, there was but one way to become a professional reader: you had to enter a religious institution. Then, when you had served your long apprenticeship satisfactorily, and passed all the tests, you could join the select ranks of the professionals. On Sunday you could read the lesson from the pulpit.

It is no wonder that the congregation was impressed by such a feat. They were in awe of the priests who could make sense of all the little squiggles on the manuscript. To them it must have seemed like magic. The priests reading the Divine Word must have appeared close to God. There was certainly a mystical dimension to the experience of listening to the lesson.

The sense that there was something sacred about reading was reinforced by all the rituals that went with the practising. Apart from the fact that it was only taught in convents and monasteries – because there were no state schools as there are now, and only a fraction of the population could attend the religious institutions – reading then was a very different process from the one we currently know.

When there were only manuscripts to work with, reading was generally a collective activity. Scholars, who were invariably training for religious orders, sat around and studied the texts, pondering over every word. They went over and over the same Latin sentences, writing them down and reading them out loud. This was the period when there were relatively few texts available and readers were supposed to know them so well that they could quote whole slabs of the Bible, for example.

This is still the case in some religious contexts. "Remember," state Lucien Febvre and Henri-Jean Martin, "even now a Muslim child of 12 is supposed to be able to recite the whole of the Koran by heart."[1] Such a child – presumably male – could have been counted among the professional readers in the scribal era.

Apart from religious institutions, there was nowhere else you could go to read, which meant that the Church had a very tight hold on information and meanings. Not only were most of the texts in code (i.e. Latin), but because reading was conducted communally and out loud, it was a fairly public activity. This ensured that there could be no disagreement, no alternative explanations or theories. The one truth of the Church could be protected, and could prevail.

Information and Control

Manuscript reading was not something that was generally done alone. Not many scholars took to their single beds to read one of those cumbersome documents by the light of a candle. Because there were few opportunities for individual readings of a text, subversive or threatening interpretations rarely arose. In this way, as we have seen, the Church was able to monitor and control what God and the saints had intended. Besides the Church had effective ways of dealing with awkward or dangerous ideas. Anyone who put them forward was encouraged to change their minds very quickly – when faced with the prospect of burning, in this world or the next.

But then the printing press made its appearance. It is not too much to declare that all hell broke loose after that.

Suddenly, the Church found that it was not in a position to supervise the reading materials or the readers. It could no longer control what was available, who had access to it, or what readers should think. That is why it was bitterly opposed to the book and went to such great lengths to put printers and their publications out of business. Permanently.

Printers then, like publishers now, wanted to sell as many books as possible. They needed to make a profit in order to experiment, to expand. Even the most conservative among them quickly concluded that there wasn't a lot of money to be made from publishing books in Latin.

---

1. Lucien Febvre & Henri-Jean Martin, 1984, *The Coming of the Book*, Verso, London, p. 22.

Of course, some people had come to believe that Latin was the only language in which texts could be written. This dead language was held to have spiritual properties, to be the ultimate in refinement and nuance; the argument that certain sophisticated thoughts could be expressed only in Latin continued for centuries.

But the printers had no problem in abandoning the supposed sensitivity of Latin for the "crude" local languages (such as English!) which had previously been spoken.

And the consequences were humungous.

The Church's Latin manuscripts (Bibles, prayer books, sermons, etc.) were translated, and were soon rolling off the printing presses in the vulgar tongues which ordinary people could understand. The only thing that stood between them and the exciting new range of information was the mechanics of reading. The magic was gone. The mystery had evaporated. The connection between reading and religion was broken. Readers were no longer revered figures. An enormous number of people who had no interest at all in being priests were keen to learn the decoding skills and to find out for themselves what was in these new books.

For the Church and its authorities, the printing press spelled disaster. The spiritual nature of reading was soon destroyed. It was no longer the Word of God expressed in an ancient language, solemnly studied in a religious setting under the supervision of the professional readers. It was a base activity where it was every man for himself, where devilish translations led individuals to come to God-knows-what conclusions about the way the world worked. It was the end of all decent standards – or so the religious authorities decreed.

Reading the new books was seen as a dangerous and disruptive activity by the religious rulers; they preached that it would undermine the order of society and the place of the Church.

They were right, of course.

Classical Arguments

Despite the steps that were taken to stop the presses, the Church could not put an end to the books that were being published. No matter how many books were burned, no matter how many printers were imprisoned, tortured, executed, the demand for books was inexhaustible, and the supply kept up with it.

There was worse to come.

It wasn't long before the printers ran out of religious texts to translate and publish. In their quest to find suitable stories that would interest their audiences, they turned to the myths, legends and histories of ancient Greece and Rome. When these were translated and made available to the public, the Church dignitaries were horrified. For these were pagan texts. They were the work of the devil. The stories had false gods, and tempted readers into wicked and wanton ways. These were not the Word of God. Quite the reverse.

This too is a fascinating development. For the pagan texts were what we have come to call the classics. They are the very stories and styles that have been revered over the past centuries. They became the foundation stones of the humanities; the high culture of the print period, the basis of good literature. They are the means by which we have enhanced communication and the sense of community.

The reception that was given to the pagan texts was not very different from many of the objections that are currently being raised about the new media. Television and computers are sometimes held to be evil, undermining the moral fibre of society, to be taking people away from the true values. Only these days it is the voice of the established humanities, rather than that of an authoritarian Church, warning that the new media will bring about the downfall of civilisation and the end of law and order.

Like the modern critics, the Church did not state its grievances in terms of self-interest. Religious dignitaries didn't go around complaining that the book was challenging their power, reducing their influence and marginalising their professional skills. Rather, the objections were all about the damage that was being done to the individual and the community. In another fascinating parallel with some of today's criticisms of the new media, the Church condemned the way information was being transformed into entertainment.

All the serious and sacred texts which dealt with the beginning of the world and the secrets of the universe were being replaced by titillating and tantalising tales from pagan times, intended to divert, entertain and amuse. Discipline would disappear, brains would go soft, honour and uprightness would be sapped by all this salacious, violent, permissive literature. There was no mention of the violence of religious texts: images of crucifixions and hell, the horror of brothers killing each other, fathers prepared to sacrifice sons, or the exploitation of women and girls.

The abuse directed at the translation of the classics was just the beginning of the lament about the drop in standards that book reading represented.

The establishment predicted gloom and doom, not to mention civil unrest, if everyone could read the dangerous books which were so freely available. Given that they couldn't stop the printing presses, nor effectively censor all that they thought subversive, the only thing to do was to prevent certain people from being able to read.

For a couple of centuries, attempts were made to keep reading confined to a small and trusted elite. Law and custom helped to keep women and the working classes (and slaves in the USA) away from books. Once more it was white, professional men who enjoyed the privileges.

The Growth of the Reading Public

The Democratisation of Reading

Over the centuries it became increasingly difficult to keep books out of the hands of "undesirables" who might get the wrong ideas from reading what they couldn't possibly understand! As Alvin Kernan says, "By the eighteenth century, printed materials were so widespread as to bring on fear of a 'literacy crisis', a crisis which was the exact opposite of ours [now] in that it involved too much rather than too little reading."[2]

One of the philosophical arguments that raged throughout the eighteenth century, and well into the nineteenth, was the extent to which the masses could be entrusted with this privilege; and the extent to which ideas and meanings could be censored, supervised, and managed by those in power.

Gradually the nature of reading was redefined. It was no longer regarded as a complex and specialist skill which only an elite could aspire to acquire. The fear that freely available books would make the population ungovernable gave way to the belief that a modern society needed its members to be literate if it was to function efficiently.

In his seminal book, *The Long Revolution*,[3] Raymond Williams traces the growth of the reading public, which he parallels with the growth of democracy. While he makes no mention of the way women were treated (another "book" waiting to be written), he does show how, among men, reading went from being a profession to a mass activity. This change is sometimes referred to as the democratisation of reading.

What are the lessons that history offers us about these shifts? What do we make of all those warnings about the dangers of reading, now? What do we think about all the predictions of the end of family values and the social fabric when – with the benefit of hindsight – we can see all the great achievements of the book era? And how should we use the information that the dreadful influences that were once associated with the book are now being transferred to the screen?

It is interesting to see the dramatic shift that took place in the nineteenth century when reading went from being a restricted practice to an open activity. While you can't actually name the date that reading ceased to be seen as a privilege, it's clear that there was a steady move in this direction throughout the nineteenth century. It reached the point where the establishment could see print in a positive light as a means of reaching the whole population, and keeping people informed about priorities and agendas. (This was the potential of propaganda – of being able to sell to the entire society.)

If the past is anything to go by, we can expect a repeat performance with the electronic media. No doubt there will be a shift away from the elite use of computer networks, and in the not so distant future, everyone will be required to be "plugged in" – to be computer literate.

Access to networks will be seen as a human right. There will be parliamentary acts which make it mandatory for everyone to be wired up, for so many hours per week, so that they too can become informed. The authorities will be able to reach – and program – the entire global village. In such circumstances, thought management will be far from science fiction.[4]

---

2. Alvin Kernan, 1990, *The Death of Literature*, Yale University Press, p. 130.
3. Raymond Williams, 1978, *The Long Revolution*, Penguin, London.
4. For further discussion, see pp. 157–60.

Print Programmed

The more the electronic media become legitimate, the more complaints there will be (in the short term) about the drop in standards and the lowered levels of literacy, both real and imagined.

This is not to deny that there are people who are illiterate who don't want to be. Nor is it to devalue the sense of shame that individuals can feel as they try to disguise their inability to decode the signs and symbols of everyday living. But it does suggest that literacy rates – and illiteracy problems – are not always what they seem.

There have always been champions of illiteracy. While never denying the difficulties and the isolation of being locked out of reading, the educator Garth Boomer[5] could also see a positive side to being unable to read. It was his belief that those who couldn't read were able to escape some of the disadvantages of "print programming". They were outside the influence of propaganda: they enjoyed a freedom to interpret the world in irreverent ways that were not moulded by the pervasiveness of print, or its particular political messages.

This might be nothing other than a romanticisation of illiteracy. But in the future, it might also be something to think about.

When almost everyone has contact with the electronic media (and this means cyberspace as well), it is possible that the media will give rise to a certain mindset, a programmed way of seeing the world. Just as the medium of print produced a particular perspective.

It's not uncommon to find contemporary commentators who believe that the emergence of a global network means just more of the same. It will be an international monolith rather than a source of cross-cultural diversity, and soon we will have just one version of the news (CNN style), rather than the multiplicity of national, local and community alternatives.

But perhaps we are really just protesting about change itself. Perhaps the idea that the medium shapes consciousness, develops mindsets, and allows those who control it to limit what we can conceive is just a form of reactionary nonsense. After all, you couldn't seriously say of the book that it reduced intellectual freedoms and possibilities – could you? Didn't the book open the way for new ideas and understandings – and aren't the electronic media another advance in the same direction?

Of course, these are some of the classic arguments and dilemmas that are associated with learning a language. It's by learning a language that every individual can communicate, and become part of the human community. But as each language sets its own pattern and gives rise to a distinctive and culturally different set of values, a paradox emerges. Language not only liberates; it also blinds us to some possibilities. The extent to which we speak the language – or the language speaks us – is a matter of great academic debate. And it is relevant for all information media.

---

5. Garth Boomer, private communication.

There are all the meanings that go with the printed form, states Hugh Mackay, and he comments on the match between the linear, regulated pattern of print, and the linear and regulated way of thinking that has been prized in Western society over the past few hundred years. This, he says, is what Marshall McLuhan meant when he declared that the medium is the message. "In other words, we learn from the character, the form, the structure of the media we use as well as from the content of the messages that flow through those media. *While we are mastering the technology of any artificial medium of communication, we can be sure that it is mastering us as well.*"[6]

Yes, print has imposed a particular form of thinking, but within that schema, and at the same time, print also expanded the range of thinking and innovating. It provided conditions for a completely new sort of creativity: the emergence of the author and the rise of fiction are two obvious examples.

The electronic media can go even further. Within the bounds that cyberspace sets, new artistic freedoms and forms will emerge. They range from computer graphics to virtual reality, and include interactive fiction. We are only beginning to glimpse the creative and intellectual possibilities of cyberspace.

These questions about liberation versus indoctrination are not an indulgence. They have been at the core of critical thinking in the print era, where the usual justification for "a good education" has been that it teaches individuals to tell the difference. They will form the foundation of new branches of knowledge, new academic disciplines, and new ways of evaluating the arts and sciences, in the new reality of cyberspace.

What we know about print skills, about reading, writing and standardisation, about propaganda, truth and education, will all be transformed in the next decade. One of the fundamental issues we will face is – are we shaping the new forms, or are we being shaped by them?

Literary Crisis

Reading the Signs

There are people who insist that reading is a defined, testable and stable skill that has remained constant over the ages. In true, standardised-print fashion, they claim that reading is reading, and that there's only one way of doing it: the way the educated, professional class has done it in the immediate past.

With this norm in mind, it then becomes logical to see any of the current changes that are taking place as a drop in standards.

No matter how you look at it, no matter whether you see print as a creative force or a curse (or computers as a blight or a bonus), there is widespread consensus that reading is not what it used to be. The issue is – what does this mean?

---

6. Hugh Mackay, 1992, "Watched Any Good Books Lately?", *Bulletin*, 11 August, p. 45, my emphasis.

The golden days of reading are generally held to be some time in the nineteenth century when universal literacy was a gift for the masses. Those were the days when, almost overnight, throughout Europe, an entire generation had the opportunity to read for the first time. And they did this without the benefit of books in the home: without parents who could read, and without any tradition of literacy.

Moreover, this goal was achieved in schools that would now be classified as sub-standard: in classes that were of mammoth proportions, by teachers who would now be considered unqualified, and by methods that would now be unequivocally condemned by the experts.

What has changed? Why is it that millions of students a century ago learnt to read under the most primitive of circumstances, while today's students, with all the advantages of specialist teachers, aids, programs and remedial reading schemes, are ostensibly illiterate in large numbers? Where did it all start to go wrong?

It is common to open a newspaper now, and find an article which claims there is a literacy crisis, and goes on to express due consternation. According to Alvin Kernan, the current controversy started in the 1960s in the USA when the term *literacy crisis* was used to indicate that the skills of reading and writing were in sharp decline. But Alvin Kernan doesn't believe everything he reads, he challenges the basis of the literacy statistics and their meaning.

When it comes to defining and measuring reading, the facts are extremely difficult to establish. A survey conducted in 1982 by the US Census Bureau concluded that 13 per cent of American adults were illiterate, writes Alvin Kernan, drawing attention to the many contradictions. Six years later, the *New York Times* did its own study and declared that nearly 10 per cent of the population in the US were functionally illiterate (that is, they couldn't read street signs, fill in forms, etc.). But at the same time that these statistics were being bandied about, the official literacy rate in the US remained firm at 99 per cent.[7]

The Reading Industry

From the claims that are made in the name of the reading industry, you would never guess that there were any doubts or inconsistencies. Billions of dollars have been spent on countless reading programs by professionals who insist that they know what reading is, and how it should be taught. The reality, however, is somewhat different.

Trying to pinpoint the precise stage at which someone moves from being illiterate to literate is not only almost impossible: it can lead to ludicrous pronouncements. My niece is a good example of how this happens.

An active participant in story reading from an early age, she proudly tried to show off her skills when she reached the school door. She read, out loud, a number of her story books. But her teacher immediately dismissed this accomplishment. She declared that the child wasn't reading – *she simply knew all the words.*

---

7. Kernan, *The Death of Literature*, p. 141.

I found this perplexing. What was reading – I asked – if it wasn't knowing, recognising and articulating the words on the page? The teacher was at a loss to explain. Because there can be no definition of reading which doesn't include *knowing the words*.

Not that I blamed the teacher entirely for such a silly statement. (I was, however, very critical of her complete dismissal of a five-year-old's competence, and confidence.) As the product of a particular training system, and a peculiar educational philosophy that has led to an awful muddle about the nature and nurture of reading, the teacher was herself something of a victim. (Not the least because, by her own definition, she had a class of non-readers to contend with.)

Not all the experts have gone along with the rules of the remedial reading school. In the 1970s a University of London expert on children's literature challenged the whole remedial reading enterprise. Margaret Meek[8] questioned the popular educational convention that the teaching of reading was best left to the professionals.

Reading without Schemes

Margaret Meek wanted to know why so many educationalists diplomatically suggested to parents that they shouldn't even try to teach their children how to read.[9] It wasn't as if the reading teachers themselves had an unblemished record. She needed to be convinced that there were grounds for the teachers' insistence that parents should keep out of the business. What evidence was there for the belief held by some teachers that parental involvement was positively harmful?

She also wanted some basic – but important – information: how many children arrived at school already able to read? When she found that no systematic statistics had been kept on readers, Margaret Meek was astonished.

No one knew. No one knew how many children learnt to read before they entered an educational institution. No one knew if or how children taught themselves, despite the importance of this information for any professional who was responsible for teaching reading. The number of children who had acquired the skills without any of the supposedly necessary aids and exercises was anybody's guess.

Some teachers didn't even bother to ask their students whether they could read: they didn't even give them a book to see what they could do. They relied on tests to tell them where the students were at. Tests don't always measure what they are meant to.

So Margaret Meek decided to remedy the situation. She undertook one of the first surveys aimed at establishing how many enrolling students were readers. She was surprised at how many there were. She was also impressed by the very different conditions under which they had learnt. It seemed that about the only thing the good readers had in common was their access to books (often in a public library) and the opportunity to read.

---

8. Margaret Meek, ongoing research, University of London, presented at staff seminars, etc.
9. Meek, research.

It was also her observation that children who had a record of illness were more likely to be readers, with asthmatic myopics being the most devoted.

Constructing a Reading Problem

The school, however, usually didn't recognise these reading abilities in its students. On the contrary, the skills that the children brought with them could actually be masked or denied by school procedures. Margaret Meek pointed out that it wasn't uncommon for the children she had identified as readers to fail the school's "reading readiness" test. And so these unfortunate young students were given exercises to improve their visual, discriminatory and phonic skills in preparation for the day when they could tackle the serious business of reading. Her explanation for this state of affairs was that children who were readers could be seeing words as a whole – combining the symbols into a unit – whereas the tests were designed to assess a student's ability to see the parts: hence the tendency for kids who could read to get lower scores than the non-readers in the class.

It was the mismanagement – and the sheer absurdity – of the assessing and teaching of reading which led Margaret Meek to declare that, while there were many kids who could read on the day of school enrolment, you couldn't assume that their achievement would last. These children could soon become non-readers. Having failed the test, been labelled as slow, and forced to do boring exercises rather than allowed to read books, it would not be surprising if they were to withdraw or quit. It was the reading regime of the classroom that would turn kids off from reading, as far as she was concerned.

This was not her only scathing criticism of the remedial industry which kept testing – and failing – more kids, who in turn would become an expanding pool of potential clients. She also drew attention to the equally absurd practice of the professionals in basing their reading techniques on the children who could not read. The industry did not study good readers to see how reading was done – it didn't even recognise this ability in many "beginning students". Instead, remedial reading programs were derived from studying children who had difficulties. To Margaret Meek, the flaws in this method were obvious. You wouldn't study someone who had a problem with walking if you wanted to teach high jumping, she quipped.

These illustrations give some indication of the unreliability of the illiteracy data. There isn't even agreement about what reading is. Is it recognising the signs on toilet doors, or is it being able to read a book? Is it knowing the symbols, or knowing the whole? Is it the ability to pronounce the words, or to understand their meaning?

Without consensus on these factors, and without any accurate means to assess them, it's understandable that definitions of literacy have often been put in the too-hard basket. This makes all statements about literacy rates suspect.

It may be that millions are convinced that literacy levels are getting lower every year, but it is hard to substantiate this belief. Much depends on how you test, what you are looking for, and what vested interest you have. Hugh Mackay has suggested that many (older) members of the community regard literacy not just as a skill but as a virtue.[10] This leads them to link changes in reading patterns with moral decline, and helps to explain the preoccupation with a so-called literacy crisis.

Another factor to be taken into account in any discussion of the extent of the problem is that most of the figures on literacy are provided by the professionals. (There are employers, parents, university lecturers, who are of the opinion that literacy is not what it used to be, but they don't usually have any hard evidence.)

The fact that it is the "experts" who provide the disturbing figures is not usually taken as an indictment of their own competence; it is not normally suggested that they are failing in their task. Nor is it seen cynically as a bid to build a bigger empire – even though the tactic is clearly successful. The more students there are who test as illiterate, as having reading difficulties, the more the profession has been rewarded with increased prominence and funding.

The role played by the profession in actually creating the problem is not usually addressed. Most commentators confine themselves to a discussion of the statistics and the puzzling discrepancies, rather than looking closely at an entire reading enterprise which is now as ritualised as reading was in the Middle Ages.

The Claim for Superiority

To the monks of the manuscript period, reading was a particular public activity. It was a spiritual act, performed out loud, and it called on years of study and practice. No wonder the professional readers of the past refused to accept that all the sinners who had their noses in a book were actually engaged in the proper process of reading. They had so little in common with the habits of the monks.

Much the same sentiments are being expressed today by the mature-age print-committed. They see reading as confined to books, as decoding the symbol system of the alphabet along with a few rules of punctuation. Because "screen reading" bears so little resemblance to their own print practice, they don't even call it reading!

On the contrary, today's print-proficient generation constantly complains that the young cannot read as they do, and that literacy levels are falling. Countless tests have been designed to support such a view.

There is no doubt that the mature members of society have acquired great print competence. And there has been more to the game than mental abilities. Even our eye movements have become amazingly efficient as we scan from left to right, to take in words, phrases, sentences at a glance. Research shows that print readers are always anticipating what comes next, which is why we can complete a sentence when words are missing. This is why speakers presenting papers, or reading from autocues, can give the right emphasis in their delivery, even though this involves seeing the whole unit and not just the parts.

But if the monks were defining and assessing the literacy levels of today's population, most of us would certainly fail the test. And we follow in the same pattern when we judge the electronic generation against our own standards.

---

10. Mackay, "Watched Any Good Books Lately?", p. 44.

We have thousands of tests for reading books, on which we do well, but virtually none for reading screens, where we might not look so good. For example, we don't have a test for "reading" (or programming) the video, because we know that the average five-year-old would have a higher score than the average person of fifty.

Recently I listened to a fourteen-year-old male student read aloud a paragraph of the newspaper. He stumbled through it. The teacher maintained that he was barely literate. Yet I was in a position to know that the student was a whiz with computers. Only the day before, concerned at the cost of my telephone bill, he had been keen to show me how to hack into the system so that I didn't have to pay for my international calls. And to do this he had to read (decode) a number of symbolic systems, not just the alphabet. He was holding a range of conceptual frameworks in his head. He was translating from one to another. He darted in and out of windows as he memorised, switched, skimmed and scanned. What does "barely literate" mean in this context? Is it the teacher or the student who needs educating?

The monks might have scoffed at the skills of the reading public, but we too are in danger of dismissing the newly acquired abilities of the electronic generation. Just because we haven't taught them what they know doesn't mean that their skills don't exist. The current literacy crisis says more about the mindset and the testing procedures of the older print generation than it does about information processing.

Racism and Other Convictions

There is another variable in this muddled area. Black students in the USA – and West Indian students in the UK, and Indigenous students in Australia – tend to be recorded as having greater problems than whites with writing and reading. They don't come up to the white, professional standard. A variety of racist "explanations" has been advanced. There have been counter-allegations that it is racist teaching and assessing (by white professionals) that is responsible for this Black, non-standard behaviour. But none of these interpretations satisfactorily accounts for the differing patterns of reading performance.

The fact that almost all the materials – books, programs, print and images – are standardised in term of white, professional males could indicate the source of some of the difficulties. But the sexist images in children's books, in textbooks and reading schemes – and the almost complete invisibility of women – don't appear to have hindered the learning of Black, West Indian, Indigenous or white girls: on just about every test girls do much better at reading than boys, despite their institutionalised denial and disparagement.

What stands out as a certainty in all this confusion is that there are many people who are ready and waiting to be convinced that things are not what they used to be, that standards are falling, and that any sign of change is evidence of decline. The media play on these fears, particularly in relation to language.

There's always a good story to be had on the literacy crisis – the spelling scandal, the grammatical gaffes, and the failure of most of the population to speak "correctly".

Many otherwise rational people seem to get drawn in. It's almost as if the creations of the old God have been replaced by a new ideal of order, perfection and sin.

But if correct language is a myth, and literacy rates a mass of contradictions, there is one aspect of reading which can be quantified and commented upon. How many books are actually published, sold and read?

Books

The Place of the Book

We begin with the number of books that are published.

There might be a panic about the passing of print and the growing popularity of the screen, but the reality is that more books are being published now than ever before. This is not the contradiction that it might first seem.

(The increase in the number of books is in stark contrast to the newspaper; each year there are fewer newspapers, and fewer readers. In Australia, the total daily circulation of metropolitan newspapers dropped from 3.5 million in 1987 by 900,000 in 1993. The number of papers dropped from nineteen to twelve in the late 1980s.[11] In the US, Bonnie Allan reports that in "1993 less than 45% of women read a newspaper every day – down from 61% in 1990."[12])

The fact that more books are being published is to some extent an indication that technology has made publication cheaper and faster. Another explanation that's given for the increase in books is that they too are part of the new demand for increased information: more of everything is called for to meet the great information needs. Yet even as the number of books increases, the importance of books declines overall. Far from being the primary medium, books are now part of multimedia – and not the most important part either.

Another interpretation of the increased output of the printing press is that of sheer desperation. We are witnessing a last-ditch stand on the part of the publishers who are trying to capture a dwindling market. According to this line of reasoning, few if any publishers are making a profit from books. And none is really sure what future needs will be. We are in an experimental phase: publishers are trying out as many titles as possible to see what will sell in the electronic world.

On just about every front, there is undeniable evidence that the book is making way for the screen. This has enormous consequences for the book industry and for the hearts and habits of society. Although more books are being published, the number of publishers is diminishing, along with book sales and book readers. In the USA surveys show that 60 per cent of adults never read a book; in the UK almost 75 per cent of the

11. Matthew Ricketson, 1994, "Reading the Market", *Flight Deck*, March, pp. 14–15.
12. Bonnie Allan, 1994, "Right Beat, Wrong Angle", *MS Magazine*, January–February, p. 93.

population buy no books at all.[13] Between 1978 and 1989, the Australia Council reported a decline in book buying in the population from 20 per cent to 11 per cent.[14]

Publishers

That publishers are not generally making a profit is easy to establish: one glance at any of the trade magazines reveals a sense of gloom and doom. As early as 1991, the *New York Times* was reporting that "people are not buying books" and that drastic downsizing was taking place within the industry.[15] In the last few years, a significant number of small, independent publishers has disappeared, and mergers have become the name of the game for multinationals. Even the multinationals have rationalised their range and titles. Some have pulled out of fiction; some have abandoned children's books (on the grounds that they can no longer compete with CD-ROMs and other forms of electronic media). At the National Book Council Summit in Melbourne (November 1993), there was growing recognition that fundamental changes would be needed in the book trade if it was going to survive. It has to be said, however, that the mood was more one of cutting back than of flooding the market with new titles.

Regardless of the reason for the book expansion, the reality is that there are now more books than we know what to do with. There have even been suggestions that the print era – the Gutenberg Age – will come to an end, not when books cease to be published, but when, in bizarre fashion, we are buried under the weight of all this surplus.

This is a reasonable assessment; it has become impossible for the major libraries of the world, for example, to continue to collect and keep all that has been published. The US Library of Congress, which holds more than 88 million items, can no longer keep pace with production; "it receives 31,000 new books and journals a day and keeps 7,000 of them". The irony is that if the library is to persist with the cherished goal of the Gutenberg era to collect all the books, then it is going to have to resort to electronic means of storing them on databases, etc. Already the new British Library has had to abandon the aim of holding all that has been printed. Eight miles of shelving is considerable, but it is insufficient for the library's traditional purpose.[16]

The fact that there are so many books doesn't necessarily imply that there is more diversity, or greater choice for readers. Indeed, some of the criticisms of concentration and conformity that have been levelled at television have also been directed at publishing. There are assertions that the medium is in fewer and fewer hands, and that the appeal is to the lowest common denominator in the quest for the widest possible audience. Such complaints are not without substance.

When Time Inc. and Warner Communications proposed to merge in 1990, James Kaplan described the workings of an empire:

---

13. Kernan, *The Death of Literature*, p. 142.
14. See Hans Hoegh Guldberg, 1990, *Books – Who Reads Them?*, Australia Council, p. 4.
15. Roger Cohen, 1991, "Weak Sales", *New York Times*, 27 May, p. 119.
16. See Kernan, *The Death of Literature*, pp. 138–40 for further discussion.

Theoretically this unprecedented corporate fusion makes it possible for a title to be published in hard cover by Little, Brown (a division of Time Inc.), featured as a Book-of-the Month Club (also owned by Time) main selection, reviewed in *Time Magazine*, issued in paperback by Warner Books, made into a motion picture by Warner Bros. and turned into a TV series by Warner Television . . . [And now included in a multimedia database.] The very concept of the book as an object to be lingered over and palpated, something that seeps into the soul, has changed. The printed, bound book itself now represents one stage in a process which turns out that most unliterary of all quantities: product.[17]

Many within the book business agree. Not only has publishing become part of a multinational empire, but the aesthetics of book production have been replaced by commodification. Says Janet Wikler, Director of Advanced Media at HarperCollins, "My generation may be the last . . . to have a strong visceral affection for books."[18]

Younger generations don't have the same reverence for the book, the same appreciation of its art and significance. For many in the Western world, it is the electronic media that are stimulating, satisfying, exciting. No doubt any survey of book sales, based on age groups, would find that the dwindling sales could be accounted for simply in terms of the habits of the younger generation, for whom the electronic media are the sources of stimulation, satisfaction and challenge.

If only this were the same for the general run of publishers!

The charge that they are no longer serving the best interests of the book is not the only criticism to be levelled at publishers in their state of reduced circumstances. Despite all the discussion about the diminished status of print and the book, there is not much evidence that publishers themselves are looking for visionary and radical new niches for their skills. On the contrary, it is not hard to find some very defensive individuals in the business who blindly believe that what they do and how they do it will continue to be in demand until well into the next century (or until their working lives come to an end). They do not seem to appreciate that their particular publishing expertise is linked with the rise and fall of print as a primary medium. But this single association with the book doesn't have to be their fate.

When the first sheets rolled off the printing presses, they were very difficult to decipher. It was the fact that the machine could produce print at all which was the fascination, and which drew people in to buy the product. Yet for decades it was probably more difficult for readers to deal with print than with written manuscripts.

Apart from the issue that the letters were not as easily identified (and there are contemporary parallels here in that the computer screen is still not as easily read as the page of a book), there was the problem that readers had to shift from a scrolling manuscript to loose sheets. Not surprisingly, they were forever losing track of their information. It was very frustrating.

---

17. James Kaplan, 1989, "Inside the Club", *New York Times Magazine*, 11 June, p. 62.
18. See Carol Robinson, 1993, "Publishing's Electronic Future", *Publishers Weekly*, 6 September, p. 46.

This unsatisfactory situation continued for more than a century, until publishers appeared on the scene and started to package information so that readers could more easily and reliably access it. Publishers introduced page numbers, and bound the pages together so that information was more efficiently organised. Then they introduced title pages and tables of contents. They divided information into chapters, they developed indexes. All these had the purpose of making the information more readily available and useable.

Publishers advertised the authors, and the genres. They came up with dust jackets, blurbs, quotes, "bytes" that would entice the reader to pick up the book, to get into the information. Publishers became the experts with print, organising, presenting, and packaging the product so that readers would not only be interested but would be able to find their way around in the medium.

There is no reason that publishers should not be taking on the same role in cyberspace. Their talents are certainly sorely needed.

In an equivalent to all those unnumbered loose sheets, we now have an undifferentiated mass of information in cyberspace. It's just calling out to be categorised, organised, packaged and presented in a way that makes it a pleasurable and satisfying prospect for today's information buyers.

To some extent Web Sites and Home Pages are starting to provide signposts and guidelines to the jumble of information. But in comparison to the sophisticated way that we are alerted to what is in books (so that we usually know at a glance whether or not we are interested), the packaging on the Internet has not got to first base.

There is a crying need for publishers to provide these services, but there seems to be no great rush on their part to take up the challenge to transform their practices for the electronic era. This suggests that it won't be the print publishers who take the leap into cyberspace. Rather, a new breed of publishers will start to do for cyber-information what the traditional publishers have done for print. That will be another stage in this information revolution.

Book Stores

Unlike literacy rates, book sales can be quantified, and internationally the sales of books are in decline. In the United States booksellers have responded by trying to find new ways to entice the public into bookshops, and to put books in homes. Some are predicting that "mass customisation" will be the solution: the major book stores will be bought up by electronic companies, and books will be custom made on the premises. Hard copy will be produced instantly from a selection of thousands or millions of titles, held in a machine no bigger than a jukebox. By the year 2000, state the supporters of this scenario, "universal texts will be downloaded to a retailer and the customer will be able to compose their own anthologies at the booksite".[19]

---

19. See Robinson, "Publishing's Electronic Future", p. 52.

Not only would the need for warehousing and distribution disappear, but customers could have everything they ever wanted delivered in one print package. So, for example, along with the latest copy of Kerry Greenwood's Phryne Fisher mystery, the reader could get lists of all the other novels, statements from the critics, information on which ones have been turned into movies, etc. And all this in print form.

If the bookshops were to go multimedia, there would be an even better offer. The consumer could get a package which contained all the Phryne Fisher mysteries, as well as interactional capacity, and detective games. Not to mention virtual reality packs and games, where the "reader" could become the beautiful Phryne Fisher from the 1920s, and live more directly than vicariously.

There are booksellers who see this setting as coldly technological and who believe that more intimate surroundings will be more conducive to book sales. Book stores "that offer a warm, engaging, responsive environment may become even more popular places to go," says literary agent, Richard Curtis.[20] With shopping becoming more of a social than a routine purchasing activity, perhaps socially centred communal bookshops will attract more people. The book store with the coffee shop, restaurant or other leisure facilities may be the model of the future.

But once multimedia becomes the norm – and there are many who believe that the time is almost here – it is likely that bookshops will reflect this development. Already in the United States there are chains such as the Tattered Cover, which has embraced the concept of an entertainment boutique, "carrying a range of product offerings that includes books, audio, video and recorded music". It's but a small step to add electronic delivery to the available options: the next stage for the Australian multimedia ABC shops?

Reading Habits

All these comments on current – and future – sales and services for books beg the question of how many of those purchased are actually being read. Jenna Price, writing in the *Sydney Morning Herald*, ventures the opinion, "not very many".[21]

She accepts that Australians could be spending more money on books than ever before (nearly $A1 billion annually). But the answer as to whether we are reading what we are buying is "probably not", she declares. (This certainly fits with the figures from surveys of reading in Western societies; see pages 57–8.)

Jenna Price confesses that she certainly doesn't read as much as she used to; and she definitely does not manage to get through many of the books she buys.

> When I was a girl, I read all day and night: torches under covers, sunrises. They lit my way as I read from dawn to midnight. These days in between the two full time jobs and three

20. Quoted in Robinson, "Publishing's Electronic Future", p. 50.
21. Jenna Price, 1994, "When One Reader's Critic Is Another's Bull", *Sydney Morning Herald*, 8 January, p. 2A.

full time children I share with my spouse, I read newspapers. That's all . . . All the adults I know love to read and rarely do it unless they are on holidays.[22]

Jenna Price is a print journalist. It's not surprising that she should read the newspapers. Nor is it surprising that she should read so much less than she used to. It is not just because she is too busy; it is also that there are alternative electronic sources available. Like everybody else, where once she used to read books, she now listens to the radio, watches a video or television – or a computer screen.

That the mature generation is buying books but not reading them is no great puzzle for Jenna Price. "Our desire to appear well read crushes underfoot the reality of having little time left to read," she says. "So we must buy books in order to get a little self respect."

The idea of having books and being a reader dies hard. As the transitional generation, reared with print but living in the electronic world, today's adults have residual characteristics. We haven't broken the habit of book buying, or of feeling secure when surrounded by books (even if they are unread, and are assumed to work on the principle of osmosis, with their very presence exerting a positive influence). We continue to create a context in which books feature prominently; but these days, they may be little more than a reassuring backdrop against which we lead our electronic existence.

Electronic Books

"We are now in the twilight of the printed word," states Richard Curtis[23] as the various possibilities – and realities – of the electronic book are outlined. Until recently, the term *electronic book* generally meant electronically delivered – an electronic version of the printed material customised in the bookshop, or delivered on the Internet, for example. But now the term is more likely to mean that it is a book in electronic form: it is "read" from a screen, not a page.

Gareth Powell, a former computer correspondent for the *Sydney Morning Herald* (where computers have a much bigger, brighter, section than the literary pages), reports that the CD disk (or electronic book) is set to replace the conventional book in educational institutions. As yet, he is not talking about images, or video, or sound. Just print: "Nothing but text. As Hamlet told Polonius [more boys!]: words, words, words." It would be simple to load on one CD-ROM disk all the books needed for one scholastic year.[24]

A school could even go into business for itself, making its own CD-ROMs. The machine for doing so costs only about $15,000, and every student could have a disk rather than lug all those heavy books to school. At the moment copyright constitutes an obstacle. But given that photocopying can be licensed, there's no reason that a scheme for electronic copying of this sort could not be devised.

---

22. Price, "When One Reader's Critic Is Another's Bull", p. 2A.
23. Quoted in Robinson, "Publishing's Electronic Future", p. 47.
24. See Gareth Powell, 1994, "The Path to Compact Learning", *Sydney Morning Herald*, 28 March, p. 45.

Then there would be no more schoolbooks. For this scenario to work, everyone would need access to a school computer and a home computer with CD-ROM disk drives, a possibility that Gareth Powell doesn't think in the realm of science fiction. "My guess is that within three years it will be as difficult to buy a personal computer without a CD-ROM disk drive as it is now to buy a computer without a hard disk drive".[25] I'd need more convincing about access and equity before I would accept this, although it is in keeping with some of the comments made by Cynthia Solomon and Seymour Papert: that if every student was given a computer, computers would be cheap enough for every student to have one.[26] This is a philosophy that Gareth Powell appears to share.

The attractions of the CD-ROM disks – print only – are obvious. But the attractions of CD-ROM multimedia – print, sound, image, etc. – are irresistible. Not only could you read what Hamlet said to Polonius on a CD-ROM multimedia disk; you could see the actors and hear them as well. You could choose, from a variety of performances, whichever production appealed to you. If it were interactive multimedia, you could make your own version of the drama. You could have Ophelia telling Hamlet what she thought of his words. Her view of the hero (and *not* from the bottom of the lake) could put Hamlet in a very different light. It would also provide a very interesting development in educational theory and practice.

Books simply cannot compete with such an information medium. They can't contain enough; they can't be accessed quickly enough; they certainly can't provide the multiplicity of possibilities, or capture the interest of students, in the way that interactive multimedia CD-ROMs can do.

The new multimedia, interactive books are already here. *Long Time, Olden Time*[27] is one of the new electronic books now available. It is an electronic version of Aboriginal history, and is a multimedia mix of interactive text, film, graphics, images and sound, on the less powerful floppy disk, as well as the CD-ROM. The reason that an Aboriginal narrative was chosen as the first in this series is that "The multimedia audio version marries beautifully with Aboriginal story telling," says Marius Coomans, the publisher of these electronic books.[28]

There are many who are excited by the imaginative potential of the electronic book and whose enthusiasm for the new experiences knows virtually no bounds. We have almost exhausted the linear textbook as a way of getting information, says Peter Kidersley, impatient to get on with the wonders of the new. "An innovative blend of

---

25. See Powell, "The Path to Compact Learning", p. 45.
26. Seymour Papert & Cynthia Solomon, 1971, quoted in Gary Stager, 1993, "Computers for Kids – Not Schools", in Irene Grasso & Margaret Fallshaw (eds), *Reflections of a Learning Community: Views on the Introduction of Laptops at MLC*, Methodist Ladies' College, Melbourne, p. 42; see also p. 120 below for further discussion.
27. *Long Time, Olden Time*, CD-ROM, Random House.
28. Quoted in Trudi McIntosh, 1993, "How to Take the Electronic Leap", *Weekend Australian*, 10–11 April, p. 4.

pictures and words creates more of a wraparound experience," he says; "interactive media meets the human need for freedom of exploration where we can search for patterns out of our existing world".[29]

Of course, traditional publishers – who may not be around much longer – are not keen on this new form. But players whose background is image rather than print are most excited about the future possibilities. They see printed books being replaced not by a lesser product but by a liberating, expanding and enriching tool.

In numbingly predictable fashion, Jack Slovol outlines his view of the advantages of multimedia history books with the comment that it would be great to have "some virtual reality techniques to place you on the battle field".[30] But this – and rationales that the major gain of virtual reality will be the ability to have cybersex – are not the sum total of the plans for the electronic book. "Even the novel will become interactive, and therefore more experimental,"[31] with hyperfiction or multimedia narrative becoming a whole new genre in which the reader also becomes the author.[32]

There will be great opportunities when all the many books of the old literary canon are available in interactional form. What fun there will be in doing the rewrites, eliminating racism and sexism, turning a few heroes into villains and victims into avengers, in changing the plots and the endings. We will be able to bring Anna Karenina and Madame Bovary back to life (and send a few men to an unpleasant death); give Jane Austen's Mr Knightley his come-uppance, bring Mrs Rochester out of the attic, have Ophelia come back from her watery grave and give more than a piece of her mind to Hamlet. Far from putting an end to the novels and plays of the Great Tradition, the possibilities of interactional fiction will introduce new life – and laughter – into the old form. But to many advocates of the fixed, standardised print text, this looks like nothing other than sacrilege.

The Shelf Life of Books

Some of us who have been reared with print have a visceral relationship with books, and feel secure surrounded by their fixed and permanent nature. We will probably never make the transition to the new medium. Apart from the fact that it seems as though there's just too much to learn, we will always feel a sense of loss and deprivation when we have to look at a screen rather than a page. Some of us will go graciously, and take early retirement, but there are others who will go down kicking and complaining that things are not what they used to be.

I am constantly amazed by the number of people who vehemently protest that they will never have anything to do with these dehumanising (devilish?) machines. Without benefit of the delights of multimedia competency, they unequivocally complain that it

---

29. Quoted in Robinson, "Publishing's Electronic Future", p. 47.
30. Quoted in Robinson, "Publishing's Electronic Future", p. 50.
31. Quoted in Robinson, "Publishing's Electronic Future", p. 50.
32. This is discussed at greater length in Chapter 4.

has been downhill all the way since the rise of the electronic media, that the book has no substitute.

"You can't take a computer to bed with you," the diehards solemnly declare and smugly rest their case.

The next generation don't get the message. Not just because they have curled up in bed with a video on a Friday evening, but because they are reading their electronic books before going to sleep.

That you can't read a computer in bed is the last-ditch argument of today's Protestant, states Maureen Ebben on the Internet. She outlines the merits of the Dataman (*sic*), the Newton and other screens that are mobile and no bigger than the average paperback. The argument isn't going to stop at this point.

"The smell of ink, one writer suggests; the crinkle of pages, suggests another," in looking for the book replacement. And "in far off laboratories of the Military Infotainment Complex . . . some scientists work on Synchronous smell-o-vision . . . while others build raised text touch screens with laterally facing windows that look and turn like pages, crinkling and sighing" as the reader/viewer moves through them. "'But the dog can't eat it' someone protests . . . and the scientists go back to their laboratories, bags of silicon kibbles over their shoulders."[33]

Anyone who thinks that what we have known as the book has a future is a long way from virtual reality. On every index, the book is making way for the new medium. This doesn't mean that it will disappear entirely; we still had oral forms (theatre included) after print was established. And print will have a significant role in multimedia. But the book won't be the *primary source of information*. Future generations will not turn to the book as the entry to knowledge, the key to another universe, the escape to another world.

Even the paper that books have been printed on is against them. The acid paper which has been used extensively in the past is now disintegrating at such a rate that it is estimated that almost half the number of books in major research libraries in the United States will soon be too fragile to handle. A report on Yale's Sterling Memorial Library states that "all book repositories are self-destructing time bombs".[34]

Paradoxically, the race is on to transfer the texts of the great books to some type of electronic format in the interest of posterity.

Those who have written books are actively aware of the changes. It's no longer a matter of making a film or television program from the book: these days the book can be made from the script.

Even books that are printed won't mean what they have in the past. They won't represent the same value. In the words of Hugh Mackay, the book could well become a souvenir of electronic experience; like a theatre program, it could be a record of the real thing.[35]

---

33. Maureen Ebben, Internet, 22 November 1992.
34. Kernan, *The Death of Literature*, p. 136.
35. Mackay, "Watched Any Good Books Lately?", p. 45.

It is understandable that with the passing of the book there should be a sense of bereavement. As one who has been constructed from books – who is a composite of so many texts, who has been layered by so many literary characters, and led by so many authorial voices – it is more than regret that I feel at being deskilled. A way of life that I thrived on has gone, and I mourn it. On occasion.

Already, I am something of a curiosity to a younger generation. They think I am magic because I can close my eyes and conjure up so many pages of print. They ask me where to find information, where to locate an appropriate quote. I recall the writer, the volume, the print on the page. "Virginia Woolf," I say. "*A Room of One's Own*. About page 42, on the right hand side and half way down." Never have they been so close to print as this, so familiar with the words. Because they cannot comprehend how it is done, they are impressed by my abilities. But these are old tricks, and they have no place in the electronic world.

Memory is a dwindling asset. There's no need to retain things in one's head when it is stored on a disk. There is no need to be able to recall quotes when the computer has a search capacity and an electronic index.

Like the monks of old who could not believe that the skills they valued and thought essential could simply be by-passed, so we too will continue to feel dismayed and deskilled. For those of us who were reared on print, much of what we do – and what we are – will no longer command respect in our society. Because so much of our attitude to print is an emotional commitment, it is unlikely that we will ever be able to acknowledge that the old is not as good as the new.

4

AUTHORS

The Birth of the Author

From Copyist to Author

Authors as we know them are a recent invention. Prior to the printing press, writing was all about copying: good writers were those (with, presumably, a "good and uniform hand") who could copy *exactly* the manuscript in front of them. Accuracy was valued above all and writers were paid copy money for their efforts, which were circulated among priests, princes, and a few professionals.

Obviously, in this process there was no place for anyone to come up with ideas of their own. Scribes who made any changes or additions were not praised for their originality; on the contrary, the term *corruption* of the text was used to refer to a manuscript that was found to be altered, as we saw in Chapter 1.

Like the photocopier today, the copier of the scribal period had to reproduce the material so that it was the same as the original. Anything less, and there was something wrong. When "writing" meant producing a lot of copies that were all exactly the same, it is not surprising that the concept of originality should have been pretty meaningless.

The ideas of authorship that we now take for granted would have puzzled the population of the medieval period. The only originality or act of creation that they appreciated was that of the deity. And that work had all been done. Right down to "In the beginning was the word . . .".

Because the Church knew all that was to be known, and because it was the keeper of the sacred texts in which what was already known was recorded, there was no tolerance for "interference". Any scribe who tried to improve on the original was seen not as a creative writer, but as a blasphemer or sinner.

The scribes worked communally and copied anonymously. (There were no title pages or acknowledgements of writers on the manuscripts.) Because they left no identifying marks, no imprint of their own individuality, there was no pressure to think about such things as creativity, imagination, stylistic features, etc.

Just as there were no concepts of authorship, there were no ideas about ownership of the written word (unless it all belonged to the Creator).

Copying was a service for the Lord – be he the Lord of all, or of the Church, the castle or the manor. Writing was considered a "mere vehicle of received ideas which were already in the public domain".[1] It was a means for expressing "truths" which had been handed down from a divine source. The Church and the sacred scribes were its custodians; they ensured that the word (and the supporting commentaries) didn't get into the wrong hands.

Outside this scholarly and religious circle, however, there were people who were using their creativity to compose material. The troubadours and minstrels performed their narratives, and often wandered from place to place entertaining audiences with their songs, satires, and ditties. If there is any starting point for the role of the author as we have come to know it in past years, it is with these "entertainers" who mocked human foibles and commented on current affairs. As long as it pleased the lords who paid them.

These travelling composers appear to have been men.

It took some time before the story-telling abilities of the minstrels and troubadours merged with the mass production of the printing press to give us the imaginative writer and the category of fiction. In the interim there were many stages in the creative process.

Once printing got under way, there was considerable pressure to find manuscripts for publication. The Latin and religious texts were soon exhausted as marketing possibilities, and printers turned to the myths and legends of ancient Greece and Rome for their materials. When they began to publish these classics in local languages, translators were needed.

While a reliable translation is a long way from being a totally imaginative creation, it has elements of originality and individuality nonetheless. Translation was a practice which started to break down the notion of writer as purely and simply a copyist. But it was still very different from the concept of creative authors who could *invent* their own narratives. Until as late as 1720, the English public was not prepared for a writer who could make things up, and create a fiction.

When Daniel Defoe published *Robinson Crusoe*, it was widely praised and accepted while it was understood to be a true story. But when it became known that the author had made it all up, he was accused of being a liar, of practising a deception and perpetrating a hoax. There was even some talk of sending him to jail for the shameless trick he had played on society.

Claiming that a story was true was an understandable rationale on the part of writers who had no sophisticated concept of fiction to fall back on, and who didn't want to be seen as "telling stories" (that is, as lying). Throughout the seventeenth and eighteenth

---

1. See Martha Woodmansee, 1984, "The Genius of Copyright: Economic and Legal Conditions of the Emergence of the Author", *Eighteenth Century Studies*, 4, p. 425.

centuries, as women took their first steps in the development of the novel, they repeatedly prefaced their books with the protestation that every word in them was true.

So Aphra Behn (1640–89) insisted that the woman whose tale she was simply relating in one of her thirteen novels,[2] and to whom so many remarkable things had happened, was personally known to the author. Delariviere Manley (1663–1724) adamantly stated that the *Secret Memoirs of Persons of Quality* that she had published had just fallen into her hands and were the real records of real persons. Similarly, Eliza Haywood (1693–1756) swore that the middle-class heroines who appeared in her work – and whose stories she told purely for moral purposes – were real people whom she had observed.[3]

For almost two centuries, there was an overlap of copyist/author. When the stories were made up, the authors claimed they were true. When the stories were already known and were held to be true – as in the case of the histories of the kings of England, or that of the Prince of Denmark – the authors took liberties with the narrative and recycled them as their own.

Many of the tales that Shakespeare worked from were already in the public domain. It was the construction he put on events, and the explanations of character that he gave, that marked his version of the "copy", and proclaimed his authorship.

Worthy of Hire

One of the first obstacles that had to be overcome in opening the way for the imaginative writer in print was to make it acceptable for authors to be paid for their creative work. The aristocracy – some of whom hand-wrote plays and odes and circulated them to their fellow courtiers – set the terms. They distinguished themselves by disdaining publication and scorning payment of any sort. As they could well afford to do.

Theirs was the true art, the high culture. And their circle was fairly exclusive. They did admit some common "composers" to their midst, but through a system of patronage which they tightly controlled. In this way, the right minstrels could be employed and the preferred authors supported. The hierarchy could be preserved. For no author then in their right mind would have turned on their patron to bite the hand that fed them.

Minstrels commented favourably on the role that their patrons played in public affairs. Authors of poems and plays were full of praise for the taste and discernment of their financial masters. There was no trade in ideas – a prospect that the aristocracy would have considered rude and vulgar.

Into this ordered world came the publisher. The printer could and would print anything that he thought would sell. He was often more of a mechanic than an aesthetic

---

2. All thirteen were written prior to Daniel Defoe's *Robinson Crusoe*, which the men of letters later universally acclaimed as the first novel.
3. For further discussion of these women writers and of women's contribution to the birth of the novel, see Dale Spender, 1986, *Mothers of the Novel: 100 Good Women Writers Before Jane Austen*, Pandora, London.

individual. He was prepared to pay for a translator or a "reader". This became one entry point into the role of the professional writer. "It was as a proof corrector rather than an author that the literary man [*sic*] made his entry into publishing," comment Lucien Febvre and Henri-Jean Martin.[4]

Taking money for literary activities of any sort was seen as engaging in a grubby business at this stage of the author evolution. It was particularly degrading for a woman who wasn't supposed to have anything to sell. Little distinction was made between selling her body and selling her mind: the fact that a woman accepted money at all branded her as a prostitute.

Of course, some people would argue that authors never really did get over the hurdle of feeling uncomfortable about financial transactions. To this day, anyone who makes a lot of money from books is seen to be something of a hack – a popular author engaged in the vulgar business of trade. Serious authors don't make big money but are expected to live in Spartan circumstances. There are some who are simply grateful to be published (academics among them).

Fortunately, this doesn't seem to apply to the film world. A writer's reputation is not lost if offered a huge sum for the film rights. Scriptwriters have often received vast sums of money even when they are not known, and have no reputation to risk.

CopyMoney to CopyRight

It wasn't until the writer could be seen as a specialist, gifted and inspired, the creator of an original work, that it became acceptable to sell the fruits of one's intellectual labour. But writers could not assume this identity until it was clearly established that they had something to sell. Literary work had to be seen as *property*.

When the first presses went into production, all the texts that had previously existed were – by default – in the public domain. That's all there was. Any printer could publish anything he could get his hands on, including that which another printer had just translated and run off on his press. No consultation with any author, alive or dead, was called for. No payment to any producer of a text was required. At a time when the only concept known was that of copying, any notions of pirating, stealing, or plagiarising were "unreal".

It is interesting to note how what is real in society changes over time. One day everything is known: a writer is a copyist, and there's nothing wrong with copying the accumulated wisdom of the community. But not so long afterwards, this is not how the world works. The reality becomes one where everything is *not* known, where there are new things to be discovered, invented and dreamt of. And a writer/copyist becomes an author, who creates these new possibilities. Once work is valued for its originality, copying must be discouraged. You can't claim the ideas of others as yours, and you can't copy them (in spoken or written form) without giving some sort of acknowledgement or remuneration.

---

4. Lucien Febvre & Henri-Jean Martin, 1984, *The Coming of the Book*, Verso, London, p. 160.

Today there is much confusion in relation to copyright, and we can appreciate that it looked equally perplexing and daunting to our print predecessors. They were challenged to find a way of constructing "intellectual property" which could allow writing to be owned, sold, licensed or rented.

In this process of change, writing has become a creative art, regarded as work. The "product" that has been made is the writer's property, in the same way that a man once owned his ox, his ass, and his wife. All that was needed to make the system function was laws to protect intellectual property rights, and an appropriate method of payment.

The principles of copyright (and now moral rights) and royalties were set up to meet these requirements. They have worked reasonably well throughout the print period. So when we feel that we cannot possibly come up with a system which provides authors with the same benefits in the electronic media, we can take some comfort from the ingenious solutions which were devised for print.

When the rules for print were being worked out, it was the grievances of authors which prompted particular reforms. Having spent a great deal of time labouring over a particular work, some of them were incensed when – in modern parlance – they saw it being ripped off.

The French playwright Molière is but one example. Ribou the bookseller got hold of a manuscript copy of Molière's *Précieuse Ridicules* and proceeded to publish it and pocket the profits. Without the modern protection of copyright, Ribou was even able to secure a "privilege" which prevented Molière from publishing it himself.

Becoming a creative writer and having something to sell were two realities that went hand in hand. One of the first steps in the direction of becoming property owners was when authors began to make up their own stories and to differentiate themselves from ancient dead or anonymous authors, who were not in need of sustenance or shelter. Having devised these new and original versions, the task for writers was to find a printer to come up with the finances.

If the printer (or bookseller: often the same person) was convinced that the work would sell well, the customary arrangement was to purchase it outright. This is usually the case today with scriptwriters: they are paid a specified sum for their work and forgo any further financial interest. Until the eighteenth century, the author, just like the painter or sculptor, was paid a one-off fee by the printer, who then owned the work. The difficulty (and the distress) of this arrangement was if the book became a huge success; then for years afterwards, authors had to watch publishers making great profits from their work while they themselves often lived in penury. This was the experience of Fanny Burney who was paid twenty pounds for the manuscript of *Evelina* in 1788. For the rest of her life she had to watch the publisher "growing fat" on her work, while she was often in "straitened circumstances".[5]

---

5. See Dale Spender, 1992, *Living by the Pen*, Athene, New York, for further discussion of payment of writers, particularly the chapter, "The Wages of Writing", pp. 237–43.

Prior to current copyright laws, publishers were not confined to reprinting a work; they could actually sell the manuscript to another publisher without any consultation with the author. This practice is not so different from the way publishing houses have "sold on" authors in all the contemporary mergers and multinational conglomerations.[6] Again, there were similarities with the art world. A painting could pass from one person to another at an ever-increasing price, while the artist could be living in poverty and benefit not one iota from the sums of money being circulated.

It is interesting to note that it was the writers who managed to get the best deal for the ongoing sale of their work. No doubt this was because they addressed the problem in writing and made use of their powers of persuasion. They wrote pamphlets, leaflets and articles arguing for their rights, which eventually resulted in copyright laws and the payment of royalties.[7]

Establishing the right to royalties – a percentage payment on all the books that were sold – was not without its difficulties. Many publishers protested and proved highly resistant to the idea of perpetually paying the author (even though the publisher was perpetually earning an income from the work). There were arguments against the principle in general, and against some authors in particular, who were seen as pretentious and as demanding inflated sums for their writing.[8]

England led the way in the shift from copy money to copyright, and from one-off payments to a recognition of the author's ongoing financial interest in their work. An Act of Queen Anne in 1710 held that it was the authors, and not the publishers or booksellers, who owned the work. They owned it for fourteen years after the sale of the manuscript. If the author was still alive at the end of that period, another fourteen years of copyright was awarded.

There was no overnight revolution in author payment. But gradually precedents were set. Intellectual property became a reality and the principles of copyright were accepted, despite the publishers' predictions that they would never make a profit. There were stages on the way. Milton, for example, in 1667 sold the manuscript of *Paradise Lost* for five pounds to Samuel Simmons; but it was agreed that if it was a success, and was reprinted, then he would get another five pounds fee. With such modifications being made, by the nineteenth century most writers had contracts which allowed them to negotiate the number of copies to be printed, and the right to bargain on any new edition.

Just as there was a golden period of readership, so too was there a golden period for authors. According to the Executive Officer of the Australian Society of Authors, Lynne

---

6. In my own case, I signed contracts with Pandora Press (Routledge) in the 1980s; without any consultation these books were sold on to Associated Book Publishers, Methuen, Thompsons, Unwin Hyman, and now HarperCollins.
7. Initially a "royalty" was paid to the Crown by a publisher for a licence to print.
8. See Febvre & Martin, *The Coming of the Book*, p. 163.

Spender, much of what authors gained has been lost in recent years. In commenting on the contracts that authors are likely to be asked to sign in the current difficult climate, she says that it is the publisher, and not the author, who gets the better deal.

> It is a rare contract where publishers spell out the relevant facts required for an author to make a sensible decision about what or how they will be paid. Usually there is a great fuss made about the advance and the royalty rate. But with no details of the print run, or the price of books, it is impossible to know what the royalty rate is actually worth. This means that at the very first and fundamental level authors are being asked to accept and to sign agreements where they cannot possibly have any idea of what it will actually mean in terms of earnings for them to have a 10%, or a 7.5% royalty. Needless to say there are no other contracts where so much is hidden and kept secret. It is unheard of with any other sort of contract for the parties not to be given the opportunity of knowing exactly what they are signing.[9]

There can be no doubt that publishers are finding themselves in increasingly difficult financial circumstances: no doubt either that many see cost cutting (rather than new forms of publishing) as the solution. But it is worth noting that there is often an assumption that the author is the cost that can be cut. In a recent call for contributions to a trade book, the publisher (who shall remain nameless) stated that "unfortunately the project would not be viable if authors were paid for their work".

As Lynne Spender has pointed out, the suggestion is rarely made that those involved in the publishing industry should be asked to forgo payment – even in hard times. Nor are they required to take their salary in six-monthly instalments as writers are required to do. If publishers were paid decently only when their judgement led to best-seller success, there would be howls of protest about injustice.

Because of the current pressure on publishing and the inequities in the system, the Australian Society of Authors has come out in favour of an up-front contract. It is an agreement to pay the writer for the work when it is done, in the same way that the typesetter, editor and publisher are paid according to the work completed, rather than on the basis of dim and distant future sales figures.

Many agree with Lynne Spender that authors these days are in a vulnerable positon. As the power and influence of print dwindles, the prestige of the author predictably declines. The introduction of an up-front contract could help to protect the writer and ensure that authors are paid first, and not last. But a feature of this new arrangement that is not usually commented on is that it brings the contracts for authors much closer to those already available for scriptwriters.

It could be that authors are already being drawn into the electronic model, as new forms of relationship beween writers and publishers evolve.

---

9. Lynne Spender, 1993, *Newsrite: Newsletter of the New South Wales Writers' Centre*, March–April, p. 4.

Plagiarism

The other side of ownership of work, or copyright, is plagiarism: "the taking and using as one's own the thoughts, writings or inventions of another".[10] But it too is a recent invention, and is also now under threat.

In the manuscript era, when the sole task of the scribes was to copy, the notion of stealing someone else's words in the process was ludicrous. Particularly if the copy was regarded as the word of God! Likewise, to the minstrels and troubadours whose life-blood was gossip, and who repeated and parodied other people's statements, the concept of plagiarism would have been absurd.

But once print came into its own, plagiarism became a real crime. It was stealing someone else's property for one's own use or profit. It was treated in the same way as the modern theft of someone else's car.

Once the rules were established, printers were not permitted to pirate texts, or to plagiarise. Nor could writers take another's literary work and pass it off as their own. The only ways you could legitimately use the writing of another person were much the same as the ways in which you could use their car: you could rent it, buy it, or borrow it with the owner's permission.

With print and the fixed text that was its product, both the property and the thefts were pretty clearly defined. Anyone who copied – who reproduced slabs of another's work in written or spoken form (because performing rights were also set up), and who tried to get credit for the work – was stealing intellectual property. They could expect to be punished for the crime.

Students, of course, were likely offenders. In the interest of impressing teachers and tutors, or of passing exams, there has been the understandable temptation to quote copiously from the greats without due acknowledgment (a polite way of putting it). One had only to present the words of the authorities as one's own clever contribution. But such practice has been definitely disallowed throughout the print era. Students – and authors, and journalists and even politicians – have been successfully prosecuted for plagiarism: they have been caught out, and fined, jailed, embarrassed or shamed.

Plagiarism, however, is set to become a crime of the past. For many reasons it is becoming increasingly unworkable in cyberspace.

In the first place, it demands that someone knows that particular passages have been plagiarised. Perhaps there was a time when this was perfectly possible. A teacher or a professor was familiar with all the sources, and could tell when a student presented work that was not their own. The work of authors was "imprinted" and recognised throughout large sections of the community, and any imposters could be readily identified.

But this assumes a limited number of stable texts. It assumes that certain individuals can know and place all the relevant information that exists. This certainly isn't the case

---

10. *Shorter Oxford English Dictionary.*

in the last years of the twentieth century as Jane Dorner, the editor of *The Electronic Author*, makes clear. Writing about ownership of a text and plagiarism in the *Times Higher Education Supplement*, she defies readers to separate her own contribution from that which she has taken from other sources.

"So whose text is this?" she asks. "Will the seams between scanned or down-loaded material and original thought be easy to spot? How obvious is it that 12 non-consecutive sentences of this article are electrocopied from articles in copyright to Times Newspapers?"[11] (Of course, no one can tell. That's the point.)

But there is more to be taken into account. The second reason that the reality of plagiarism is under attack is that it depends on the concept of originality. This concept is being seriously questioned in the new literary theories.

"As any self-respecting deconstructionist will tell you," comments Jane Dorner, "any text is the product of other texts." For centuries we learned to accept the originality of the author as real, but recent questioning begins to make the foundations look somewhat shaky. It could be argued that the only way authors get ideas is from other people's ideas. Writers develop styles from studying other writers' styles. They adapt, select, synthesise, from all that has gone before, and from all that surrounds them. It makes the theft of "original" ideas something of a nonsense.

This is not the end of the challenge, however. Deconstructionists have not only refuted the idea that an author can own words; they have also queried the principle that words contain meaning. Meanings are in people, not in the symbols, they argue; in reader response theory they grant the readership the role of deciding what is meant. All this has led Alvin Kernan to declare that plagiarism, along with the author, is now dead.

Not only is it impossible to pass off someone else's words as your own if no one is entitled to property rights to them; but if the words don't mean anything anyway, you can't be accused of stealing someone else's ideas!

We must recognise that all this is already happening before we even begin to tackle the directly erosive processes of the electronic media.

Whereas print provided a fixed text, in which the writer's word on the page remained the same no matter how many copies were made, the electronic media and on-line and interactive facilities are ushering in an entirely different set of conditions.

"Seven seconds," writes Ian Verrender, "That's all it takes now to transmit every word ever written in the history of mankind from one end of the globe to the other." (If women's works are included, it may take a little longer.)

He adds, "at millions of points along this new 'information superhighway', a myriad of users can access these words, change them, reshape them, and retransmit them."[12] While he is concerned about the implications of this for freedom, and the opportunities it represents for future control, Ian Verrender also wants to draw attention to the fact that such plagiarism and theft impose a high toll on business. "With the new technology,

---

11. Jane Dorner, 1993, "Whose Text Is It Anyway?", *Times Higher Education Supplement*, 23 April.
12. Ian Verrender, 1994, "The Thought Police", *Sydney Morning Herald Spectrum*, 26 March, p. 4A.

the opportunities for pirating are almost unlimited," he writes. "As a result, the importance of copyright and the ownership of intellectual property has been elevated to the fore within the international trade community."

At the Uruguay round of GATT negotiations, there was belated recognition that intellectual property is now more valuable than agriculture. It is probably also more threatened in the current climate. But you would never guess this from the deliberations. We can expect, however, to see some changes made now that the United States has estimated that the illegal counterfeiting of intellectual property – which includes print, music, software and trademark violations or image plagiarism – is now worth $US60 billion.[13] Those charged with the responsibility of collecting payments for copying will have to come up with some way of keeping check on who uses what. It is a challenge that Michael Fraser, director of the Australian Copyright Agency, has taken up on behalf of Australian authors. But the solution is still elusive.

At the moment, it is just impossible to keep track of what is being used by whom, and for what purpose. Particularly on the net. Jane Dorner outlines the ways in which technology is transforming the particular boundaries of authorship (those related to art and music are equally being transformed).

She takes the example of the researcher – which, along with the term *user*, could soon replace *author* – who is plugged in via computer to the world's databases. Information consisting of sound, images, print, graphics, etc. can effortlessly pour in from these international sources and appear on the user's own computer. "At that point, control of the text passes into the user's hands," Jane Dorner writes; from then on, everything is up for grabs.

Of course it doesn't feel like stealing to appropriate the words/images on your own screen. "The cut and paste power of the word-processor encourages the illusion of ownership," Jane Dorner says. With a few simple interactions the distinctive contribution of the user can be imposed and the work can even look like your own. How is anyone to know the difference? There are no tell-tale marks of alteration to provide any clues.

Besides, we accept that you can tape radio and TV programs in your own home, and there's nothing to stop you from editing, adding to, and modifying the tape. So why not download to your computer, copy on to your own disk, and fix the text any way you want?

So pronouncements about the death of plagiarism don't seem to be premature. The old reality cannot be accommodated in the new medium. We have to abandon the notions of ownership that we have had for the past few centuries and find a new and as yet unknown way of establishing property rights in the global village.

---

13. The protest by music-makers outside Parliament House, Canberra, in April 1995 was to make the point that it is not illegal to *transmit* on the Internet; and that anyone sending the music of a band down the line could reach potentially 30 million listeners without paying a penny to the artists involved.

The Death of the Author

The Artist, the Garret, and the Muse

Not so long ago, writers were diligent, detailed and devout copiers of the sacred texts: by the nineteenth century they had become original authors, with special gifts, the shapers of social values – and inspired by the Muse. This change in role and reality is little short of extraordinary.

One of the starting points for the construction of the author was the notion of individuality. There could be no original work, with property rights, without the individuals who thought it all up. They then got paid for their efforts. Just as the sixteenth-century religious authorities predicted after the printing presses started rolling, the drift throughout the print era has been towards individualism in Western societies. Individualism is the ripest of conditions for the emergence of the author.

In the scribal era, writers worked collectively and anonymously. Everyone was required to adopt a uniform style so that joins between one scribe's efforts and another's were seamless. At the height of the manuscript period, when multiple copies of a particular text were needed, it was not uncommon to find a team of scribes, each with their own specific responsibility. One copyist, for example, copied the same page, over and over again; another illuminated certain letters of the alphabet, etc. The art of copying was to make sure that the manuscript was the same throughout, and – in most cases – betrayed no signs of an individual hand.

Anonymity, however, was not the style of the new printers. Prototypes of the capitalist, the printers were interested in profits, and in advertising their product. They started to put their own names and addresses at the front of the book. As they obtained the services of translators and authors, they began to promote these individuals as well, and to encourage new forms of celebrity.

Elizabeth Eisenstein has suggested that it was the very nature of print itself which seduced certain people into becoming authors. Print offered the possibility of immortality, she argues, and therefore prompted a drive for fame. "The wish to see one's work in print (fixed forever with one's name in card files and anthologies) is different from the desire to pen lines that could never be fixed in a permanent form [and which] might be lost forever".[14]

As the idea of the individual became increasingly established in the collective mind, authors (and artists of all kinds) began to sign their work, to claim the unique nature of their contribution. And, state Lucien Febvre and Henri-Jean Martin, "Contemporary writers who had their names attached to hundreds and thousands of copies of their work became conscious of their individual reputations":[15] they had something of value to guard and protect. Some became very precious about their achievement.

---

14. Elizabeth Eisenstein, 1983, *The Printing Revolution in Early Modern Europe*, Cambridge University Press, p. 84.
15. Febvre & Martin, *The Coming of the Book*, p. 261.

"So little by little the profession of author was created. Slowly the author came to recognise and obtain recognition for his right to profit from his work and to have rights over his product."[16] The more the authors were able to negotiate their own incomes with publishers, through sales, the further they moved from the ties which for so long had bound artists to private and aristocratic patrons.[17] They were no longer obliged to flatter powerful figures in the hope of support: they could be independently financed and independently minded.

Not that Lucien Febvre and Henri-Jean Martin think that the system that evolved was optimal. "Now that he had a share in the profits," they write (and it could be that their use of "he" represents accuracy rather than bias), the pressure was on the author "to try and produce the best seller; and so he had to aim for the widest possible readership. In the end this perhaps served to encourage hack writing rather than work of the highest quality."[18]

Today, when one of the most scathing criticisms levelled at television is that the entire industry exists to make a profit (rather than quality programs) and that in its appeal to the widest possible audience it descends to the lowest common denominator, it is fascinating to note that this criticism can also be directed at books and authors. The print period is commonly held up as the reference point for quality: where, until the grubby commercial practices of the electronic media, the notion of art for art's sake prevailed. The author was elevated to a position of authority and privilege, above the sordid consideration of sales.

Throughout the eighteenth century, authors became more visible and more respected. The works they created began to be regarded as a legitimate way of knowing; they were engaged in an enterprise that held up a mirror to society, and provided insights into the nature of individuals, the community, and the meaning of existence. The role of the author was increasingly romanticised. Authors were seen as special, as having discernment and powers of expression that distinguished them from ordinary mortals. Writers could make "lovelier things and truer things than the rational mind can discover in its laboratories or with its computers".[19]

To explain this romantic rise of the author in terms of the lure of fame and fortune is to tell only half the story. While the prospect of being known and celebrated might have attracted men to the writing cause, it had little to tempt women.

At the same time that men were signing their names to thousands of copies of a text, and claiming identity, reputation, and a privileged place in the world, women were renouncing public acclaim and profits. There was a pervasive sexual double standard when it came to the creation of the author, and women had to face the fact that, far from enhancing their lives, the achievement of a public reputation (and the

---

16. Febvre & Martin, *The Coming of the Book*, p. 166.
17. See section on Samuel Johnson and his struggle with aristocratic patrons and the dictionary, pp. 22–4.
18. Febvre & Martin, *The Coming of the Book*, p. 166.
19. Alvin Kernan, 1990, *The Death of Literature*, Yale University Press, p. 190.

acceptance of money) spelt social death. To attract attention to oneself (to "sell" oneself) was to step outside the bounds of decorous behaviour. That is why, until well into the nineteenth century, many women were trying to avoid being known by using pseudonyms or writing anonymously.

Women who were public figures invariably defended their womanliness and justified their writing. As Janet Todd points out in relation to the professed reluctance of some of the early women writers, "Elizabeth Inchbald complained of being forced to authorship, Charlotte Smith bitterly resented her dependence on her pen, and Sarah Fielding claimed she wrote because of 'Distress' in her circumstances."[20] They justified themselves so that people would not think too badly of them.

This was not the only gender difference in the construction of authorship. As the image of the author as a creative, original and special and talented individual took shape in the eighteenth and nineteenth centuries, the need to name the source of inspiration also arose. If divine inspiration wasn't the origin of the creative impulse (as it had been during the scribal era), what was it that moved authors to produce their works of art?

Enter the Muse.

According to the *Shorter Oxford English Dictionary* (for what a dictionary definition is worth!) the Muse was one of nine sister-goddesses, the offspring of Zeus and Mnemosyne (Memory). They were regarded as the inspirers of learning and the arts, especially of poetry, and represented as young and beautiful virgins. Whatever version of the Muse was used in the nineteenth century, she was always unambiguously female and she was frequently called upon to help the male artist with the act of creation. At a symbolic level, then, the very process of authorship was represented as male.[21]

Women were kept out of authorship by the threat of gaining – and thereby losing – their reputation: they were deprived of education and the opportunity to learn to write: they were denied symbolic entry to the creative process. Women were also kept out of authorship by the simple practice of men appropriating women's writing and passing it off as their own work. When you put all these factors together, along with the romanticised concept that the most dedicated writers could be found living bohemian lives in a garret, where all else was sacrificed to art, you can see why men thought they owned authorship. Women felt that they were not supposed to trespass into the area; women often lacked the conviction that they could be authors in their own right.

There can be no better illustration of the male right to the territory of authorship than the reply that the poet laureate, Robert Southey (who has heard of *him* – except in

20. Janet Todd, 1984, *A Dictionary of British and American Women Writers, 1660–1800*, Methuen, London, p. 8.
21. No doubt there were women artists who could have been inspired by a female Muse, but this was not the intention of the symbolism. For further discussion about the maleness of the creative act, see Sandra M. Gilbert & Susan Gubar, *The Madwoman in the Attic: The Woman Writer in the Nineteenth-century Literary Imagination*, Yale University Press.

this context?), gave to Charlotte Brontë when she sought advice about her prospects as a writer: "Literature cannot be the business of a woman's life and it ought not to be. The more she is engaged in proper duties, the less leisure she will have for it, even as an accomplishment and a recreation."[22]

The dominant image of the gifted writer throughout the period has been male. (Women may have been accepted as scribblers, they may have even been popular, but their work, and their novels, were virtually never placed in the same category as that of their male counterparts. Needless to say, this has nothing to do with quality.) Women's role was to serve male genius, either as muse or maidservant. Many of the books that have come out of the dynamic feminist scholarship of the past decades have addressed this issue of male appropriation of the creative realm, and the devaluation and dismissal of women's literary contribution.

This probably helps to account for the fact that women have not been seen to put up the same defence of the privilege of authorship. Women have never had the same vested interest in it, so they are not so distressed at its deconstruction.

Indeed, Alvin Kernan accuses feminists of assisting in the attempt to destroy the concept of past eras that the author was the great, gifted authority. "The author, that central pillar of romantic and modern literature, whose creative imagination was once considered the validating source of all art, was finally sent to the guillotine," he writes bitterly, as he singles out two men who have participated in the execution. "Michel Foucault, 'What is an Author?' and Roland Barthes, 'The Death of the Author' insist that the author is only an historical idea, as Barthes says, 'formulated by and appropriate to the social beliefs of democratic, capitalist society with its emphasis on the individual'."[23]

It doesn't seem to matter what the source: the consensus is that, in the terms that we have known, the author is now dead. The concept of the professional writer, the person with special talents and a wonderful way with words, who publishes in print, is going the same way as the professional reader.

There is as much concern about the spread of democratisation to writing as there once was about its spread to reading.

It remains to be seen whether it will take as long for the idea that everyone can be an author (and be required by Education Acts to become so) as it took for the idea that everyone could become a reader to gain acceptance. But it does seem likely that, just as readership increasingly empowered the masses and made them more independent learners and assessors of information, so too does electronic authorship increasingly have the potential to empower those who have access to it.

There are gains to focus on along with the pain. For those of us who have been reared with print – and who have been authors – in the context where authorship has

---

22. See Dale Spender, 1989, *The Writing or the Sex?*, Athene, New York, p. 25; also includes further discussion of the discouragement of women.
23. Kernan, *The Death of Literature*, p. 73.

been highly skilled and privileged, there is a sense of loss as the authority of the author is transformed. It is a form of death, and mourning is called for. Some of the time.

The Democratisation of Authorship

With the death of the print author, will there be a role for the professional author? With the democratisation of authorship, where (on the net) everyone can have a turn, will there still be a place for some people to earn a living from writing? Or will the whole enterprise of authorship lose its status and credibility, as professional reading did once it ceased to be a specialist skill and was expected of everyone?

No one can make definitive predictions for a distant future; too much is changing too quickly. But within the foreseeable future there is obviously a role for the writer: as the creator of characters and storylines in CD-ROMs and narrative games; as the author of electronic books; and as a script writer on a multimedia team. These are only the initial possibilities.

It has to be acknowledged, however, that the days when the author could "write" simply in the medium of print are numbered. There will be a very limited audience for the single medium: probably only those of the present print-proficient generation who don't make the transition. Multimedia will be taken for granted by the electronic initiates. They will want words and sounds and images. Their narratives will be less like a book, more like a film.

This is one reason that authors will be more likely to work on a team than as individuals.

In the professional journal of the British Society of Authors, *The Electronic Author*, Michael Hiley describes the nature of the new writing task and outlines some of the new skills that the author of multimedia will require: "The business of being an author is going to be much more that of a team leader, and the new multimedia publishing will be more like film production, using multiple, overlapping techniques."[24]

Of course, to literary critics like Alvin Kernan, this scenario is catastrophic. To think that there is any progress or improvement as traditional book authors move in the direction of becoming scriptwriters is to him abhorrent.

> The author has been a major figure in literature, the person who knows something and writes it, but television [and multimedia] like the moving picture earlier, demotes authors to scriptwriters, or treatment editors, and buries them even more than films, among the host of other production credits in which only the star actor and much less frequently the producer and director, stand out.[25]

And, adds Alvin Kernan, bitterly, "In fifty years television has not made a single author famous."[26]

---

24. See p. 89 for further discussion.
25. Kernan, *The Death of Literature*, p. 148.
26. This statement is more applicable to United States television than British, where Dennis Potter has become famous for his television scripts; though admittedly he is the only one I can think of.

Probably no one will dispute that authors of the future will work collaboratively, rather than as lone creative individuals. (If anyone will become the professional, elevated to star and sage status and seen to be special and gifted, it's probably the director: already there is a cult emerging in relation to this role. There is some competition from the stars, who of course enjoy celebrity status, though they are not always credited with being creative or gifted; more emphasis is generally placed on their private lives than on their talent, and many state publicly that they want to be taken seriously.)

Creative authors and members of the literary community could well protest when they find themselves being aligned with scriptwriters; but scriptwriting is a form of authorship that has emerged with the electronic media and which could be the form that becomes the model for cyberlit. After all, scriptwriters work as a team, particularly on a script for a soap, a documentary, or even a film.

Scriptwriting can be both collaborative (where the team assumes the responsibility for a series) and individual (where one person does the script for one segment or episode). Who has done what is noted in the credits; there are even acknowledgments that a series or a screenplay, for example, is based on the idea of an individual. But most of the time, most of the authors are anonymous. More often than not, they are paid a one-off fee, without regard to the future success or use of their contribution.

It seems clear that in such a context the author is less likely to be seen as a source of inspiration and originality with rights to individual ownership of the work. (Not that this means that scriptwriters are not creative: on the contrary, they are often seen to be very dynamic ideas people; but they aren't seen to own the ideas or the scripts that they contribute to, particularly not in television.) In the future, authors are going to have fewer opportunities to become the sort of celebrities they have been in the print era.

There is evidence that some publishers are relishing these new developments. They are taking advantage of the changes to reduce the role, status, and remuneration of creative writers. Where deals are being done in relation to CD-ROMs, some publishers refer to the computer programmers as the authors! This leaves the story-writer in the invidious position of being relegated to "text provider", a mere word-enterer in a multimedia environment. Such a person doesn't have to be paid artistic rates for an original contribution.

It's easy to see why publishers might be tempted to make the writer just another worker on the production team. As such, they have no claim to copyright or royalties (a publisher's dream!). But there's no reason to think that this arrangement will become enshrined or will persist. Indeed, traditional print-publishers are themselves facing such difficulties in managing the transition to multimedia that it is just as likely that they will be the ones who don't make it into cyberspace.

This does not change the fact, however, that authors are going to have to learn some new tricks if they want to make a living in the electronic world.

At this stage, it seems that there will be a move towards the more collective and anonymous contribution that we currently associate with scriptwriting. In some ways, this new form of authorship has more in common with the old scribal period than it does with the patterns of print. A script team meeting to work on a soap, for example, has more to link it with the monks working together in the scriptoria than with the individual who worked alone in the garret to provide a unique and immortal work. Apart from the fact that contemporary scriptwriters are as concerned as the scribes of the past that the seams don't show in their collaborative efforts, there is also the issue of an "enduring work" which separates the scriptwriters from the print author.

Print provided a fixed medium which encouraged authors to think in terms of permanence and posterity, a legacy which could be handed down through the generations. It is doubtful whether writers working in the electronic media see their role in this light. The transitory, ephemeral nature of the electronic media is one of its most striking features.

Clearly there is going to be (heated) debate on the issue of these changes in authorship, and whether they are a good or bad thing. On one side, we have those who are – to put it bluntly – being deskilled. They are having their privilege and persuasive powers undermined as they, and their work, cease to be valued in the way that they have been in past centuries. They feel quite strongly that their passing is society's loss.

On the other side is a new breed of authors. They can function in the electronic media. They embrace it; they are enthusiastic about its amazing possibilities. They see the disk as the symbol of a new world, and are not at all threatened by the prospect of being deskilled. They see the birth of innovative and widespread forms of authorship as a bonus for traditional writers, and society's gain as well.

"It is the most exciting development in my nearly 40 years as a professional writer," writes Colin Haynes in a letter to the editor of *The Electronic Author*. "We authors already involved in electronic publishing are pleased to share information because we are so enthusiastic about our medium." He adds a note of warning: "Writers lost control of print directly it was invented. I hope we have learned enough over the past 500 years to ensure that . . . we don't make the same mistakes now that this new technology gives us real publishing muscle direct from our disk."[27]

Many authors who are at ease with the electronic media welcome the support that software can give. In the last few years – for the first time – studies have actually been done to work out how people write (as distinct from the myths of how people have been taught to write, or how the community thinks writing is done.) Yet another blow has been delivered to the construct of "originality". What has emerged is "that the largest part of a writer's time is not spent in *creating* text but in forming ideas, planning and revising."[28]

This is where the computer can provide invaluable assistance. If – as Mike Sharples suggests – "Most writers do not go through a simple sequence of planning, drafting, and revising but instead they move back and forward between ideas, plans

---

27. Colin Haynes, 1993, letter, *The Electronic Author*, 2, p. 2.
28. Mike Sharples, "Beyond the Word Processor", *The Electronic Author*, 1, p. 12.

and text," then software can readily be designed to facilitate this technique.

> Research on the writing process has inspired a new generation of "writing environments" [which] aim to assist a writer through the process of writing, from capturing ideas to producing a finished text. One of the first of these was the XeroxNoteCards which provides the computer equivalent of file cards. A writer can brainstorm ideas, write each one onto a simulated file card, and then link them together graphically on the screen.[29]

According to Mike Sharples, all this helps to make creativity visible. Of course, one person's delight is another person's despair. There will be many print authors who won't want creativity to be treated in this practical manner; it takes all the magic out of it. This is the same complaint that was made about reading when it ceased to be a gift, a talent, a profession, and became something that anyone could do with the right materials. Given the development of the reading industry and the vast aids that support remedial reading (Chapter 3), the possibility must be entertained that there will be a similar growth in writing; it could soon be programmed and prescribed to an extent that makes reading technology look simple in comparison.

Removing the mystery and magic from the writing process is something that has been going on for years. How else to explain the number of how-to books on authorship that have appeared: and how else to explain the increasing availability of classes on creative writing in higher education? Ten years ago they were almost unheard of – and when they were, they were often mocked by serious writers who dismissed as preposterous the idea that you could mechanically teach the special skills of authorship. You could not teach that elusive quality of originality, talent, aesthetic judgement.

In keeping with some of the research that has been done (and the marketing strategies of the industry), the next giant step in the writing business could well be in the form of software. What will the consequences be?

Will students need teachers when they can have computer programs which help them write in all sorts of new ways? When they can be networked to other writers in their own class, or all over the world, and work as a cyberteam? It is not too much to suggest that there will soon be thousands of authors, perhaps even millions of network users, who will be debating the pros and cons of various writing packages through the medium of an electronic conference. Then we can say that authorship has been well and truly transformed.

Currently, authorship is in a transitional stage; it is poised between print practices and screen skills. There are some writers who represent this transition, who have a foot in each camp. Established novelists like Jeanette Winterson, Helen Hodgman and Helen Garner for example, write film scripts,[30] as do authors such as Anne Deveson and David Williamson. These writers could undoubtedly – and with great confidence

---

29. Sharples, "Beyond the Word Processor", p. 12.
30. For further discussion on the way women authors are making the transition from publishing to performing, see Dale Spender, 1991, *Heroines*, Penguin, pp. 1–28.

– make the shift to computer-assisted writing and to writing multimedia packages for other writers and users.

While there are many members of the literary establishment who are understandably critical of these latest developments, the message which comes from media departments is much more positive. Jan Bruck, for example, teaches media studies, writes academic articles, and makes documentary videos; to him there is no contest when it comes to assessing and comparing the old forms of authorship with the new. He starts with the mass media of radio and television and shows how they are creative – and critical – forms. They allow many more people to participate than was the case with the old literature, which had an exclusive and elitist image and made most people outsiders.

To Jan Bruck, the different views, values and vehicles of the new media cannot simply be read as deficient or labelled as a drop in standards. What he sees is:

> The emergence of a different set of creative minds which gathers round the electronic media – journalists, TV producers, media critics, talk show hosts, and the many agents and stars of mass culture who are now displacing literary figures, ie. philosophers, writers and artists, as the representative public voice.[31]

This is the democratisation of authorship. As far as Jan Bruck is concerned, it is an extension of the information franchise. "The print media were the backbone of industrial society," he writes; "the electronic media are the backbone of the post industrial society, and the language of the majority." He goes on to illustrate the ways in which they provide a forum for most people to express their political interest.

In "breaking down the traditional authority of authorship, speeding up the process of information exchange and giving access to broad sections of the population," he argues, the electronic media are doing much more than print to create "a new public sphere in which the majority of people can participate".

In his appraisal of authorship, Jan Bruck hasn't really included the implications of the computer screen and multimedia.

> In contrast to the authority of the writer which is based on the notion of exclusivity of knowledge, the authority of those who appear on TV as presenters or stars is based on the idea of a knowledge which is shared by everyone.

How often do you hear the disparaging – but erroneous – comment that it doesn't take any skills to be a TV star, talkback host, or writer of a soap? Anybody can do it! This is really just another way of saying that it is common knowledge rather than exclusive knowledge which is on the agenda.

The premise that authorship is being democratised is based in part on these claims that the audio-visual language of the screen doesn't call for the ability to decode an

---

31. Jan Bruck, 1991, "Writing in the Electronic Age", *Media Information Australia*, 61, p. 51.

alphabet to be understood by everyone. The notion of the egalitarian nature of author-ship will be extended even further as networks and on-line multimedia with interactive capacity are included in the range of forums to which everyone has access. (We should never lose sight of the fact that, even as we talk of the democratisation of authorship, we are speaking of First World societies and highly privileged communities.) If con-nected to cyberspace, then every user has the potential to be an author, to put informa-tion on the net which can be transmitted to the world. There will be no teacher, editor, publisher or bookseller to vet or validate what goes public; it could soon mean the end of the rejection slip and the demise of the gatekeeper.

It will be the age of mass authorship in the same way as the nineteenth century was the era of mass readership. But until there is public access to the networks, all this is more in the realms of future potential than present reality.

In the current climate, authors face pressure to turn their attention to electronic, soft-ware, and multimedia opportunities. There is understandable apprehension. However, this should not be used as a means to dismiss these forms of authorship as a lesser task, or as a diminished skill. To be familiar with the new technologies is to appreciate the challenge of – and the potential for – telling stories and playing word and image games (the next two sections deal with these issues). It would be very difficult to construct a convincing argument that the electronic media present professional writers with fewer – rather than more – opportunities for imaginative work.

But when the field is opened up to so many more people, you won't need an agent, a publisher or a promotions person to get your work made public. There won't even be a need for vanity publishers when we can all have access to the Internet.

To draw a parallel with reading, it wasn't that the professional readers lost their ability, or that books reduced their scope; but that everyone was allowed to get in on their act. In a great imaginative leap, reading was assumed to be a human skill – a human right – and it became the conventional wisdom that reading could be acquired by *all* members of the community.

So should authorship be seen in such a light. We are now on course to see compo-sition as a human skill, the production of information as a human right, within the range of all individuals and not just limited as a privilege to the professional few.

No doubt some professional readers from the Middle Ages adapted to the new demands of books; they didn't all just fade away. Obviously, contemporary professional authors have some choices in this transitional period. It would be a mistake for them to think that they have nothing special to offer in the new democratic context in which so many can be empowered with authorship.

Print-based authors can move towards the new frame of values, and join the many bright and emerging authors of the electronic generation. Everywhere the print professionals are being urged to become part of the new medium, to enter the new age, to bring their experience and expertise into cyberspace, rather than cling to past goals and glories. The professional literature presents information on the possibilities of reshaping authorial skills in line with these new demands.

In an age of mass authorship, there can still be a place for those who earn their living by their disks.

Michael Hiley turns to an earlier example of artistic evolution to make this point and to encourage authors to make the change.[32] Miniature painters and wood engravers were the experts at creating likenesses, he says, until the advances of technology came up with photography and put them out of business. But the miniaturists and engravers made the shift; they adapted their craft to the new medium, and many became photographers. Michael Hiley throws down the gauntlet to contemporary authors to make the transition: to become part of the mass movement or fade away.

Telling Stories

Professional authors are confronting two fundamental areas of change. The first is competency in cyberspace. Whereas authors once had to be print-proficient, they will now have to be computer-proficient. At the level of technical skill, there's much more involved in utilising the capacity of the computer than there was with a pen – although, as the next generation's ease with technology demonstrates, it's not all that difficult and to some extent depends on what you grow up with.[33]

But apart from the technical skills which have to be acquired, there is another area which is of enormous significance to authors, particularly the writers of fiction: the medium makes a difference to the message.

It is widely accepted that a particular medium gives rise to a particular way of looking at the world, that it facilitates certain mindsets, concepts and art forms. Print, with its regularity and standardised format, helped to produce an intellectual climate in which order, laws, and linearity were the foundations for understanding in the sciences and arts. As we look back on the print period, we can begin to identify the connections between print, the Industrial Revolution, and literary forms, Jay David Bolton, writing in the well-established journal *Computers and Composition*, makes a link "between the relentless linearity of the Victorian novel and the technology of the rail".[34]

But if the medium does help to shape the form, what will the new messages be? We can see how print gave rise to the novel, but what sort of art and literary forms will emerge from the computer and interactive multimedia? What will "stories" be in the future?

The study of the mindsets of cyberspace and of the mental massaging of screen and disk is as yet in its infancy. Nevertheless, it is possible to identify the major differences between print and the electronic. Whereas print was fixed, regulated, mono-dimensional, and linear in its characteristics, the electronic media are moving, unpredictable, multiple, and non-sequential.

---

32. Michael Hiley, 1993, "Writing for Multimedia", *The Electronic Author*, 2, p. 8.
33. This is discussed at greater length in Chapter 5.
34. Jay David Bolton, 1993, "Alone and Together in the Electronic Bazaar", *Computers and Composition*, 10, 2, April, p. 13.

Stories and everything else, including academic articles, will be markedly different, as they are authored in the framework of the new media. A fascinating and fruitful area of study in the next decade would be to monitor this unique transition period, and to trace the emergence of the changing genres and forms.

Current interest seems to centre on the way writing (and perhaps reading) changes as authors move from page to screen. Questions about whether we write more or less, longer or shorter sentences, etc. are to be found scattered throughout the professional literature; but they invariably betray the origins of the questioner. They are still print-predicated. They transfer the print mindset to the electronic media. They refer to the computer as a word processor, something like a fancy typewriter, which is a more sophisticated device for producing the print or hard-copy version that is regarded as the real thing. But the computer is not just a tool. Like print, it is a medium with its own ability to shape consciousness, and to restructure society: from the workplace to the realm of private relaxation and entertainment. It makes a fundamental difference to what we know, how we know it, what makes sense, and how we express it.

For these reasons, traditional authors can either become *au fait* with its possibilities and be part of the production of new forms, or be passed over. While there is a time-frame in which the new skills can be acquired, it will not be as long as the gestation period that was associated with print.

The printing press got under way in 1450. It was two hundred years before the novel was born. Its birth was not quite as late as 1720, the date of Daniel Defoe's *Robinson Crusoe*, which the male literati have accorded the status of the first novel. Women's contributions did not count in their scheme of things but it was the case that Aphra Behn (1640–89) published thirteen novels in her lifetime which, along with her seventeen performed plays (and translations and poems) made her the most popular, prolific and well-paid writer of the Restoration period.[35]

But while Aphra Behn's thirteen fictions represented a radically new genre, they didn't spring from nowhere. Throughout the seventeenth century there were develop-ments in story-telling which ranged from the emergence of diaries, and the early biographies and autobiographies (as individual experience began to be seen as real and worthy), through to the initial epistolary novels. All of these paved the way for the dominant form of story-telling for the next few centuries.

There have always been story-tellers of course. But within the oral tradition there are limits to what a story-teller can do. Performance plays a vital role, for example, as does the ability to handle an audience. Although story-telling flourished prior to print, it was very different from the form that became the popular three-decker novel of the nineteenth century.

---

35. For further discussion about Aphra Behn and women's role in the development of the novel, see Spender, *Mothers of the Novel*.

Writing (and then print) did something absolutely extraordinary: turned the story into a concrete thing that you could pick up and examine in detail, put on a shelf and come back to, mass produce and sell. It allowed story-tellers to work "off-line" and develop elaborate traps and delights for their audiences, distort time and space, speak with different voices, and develop what we now recognise as "plots".[36]

One of the high points of achievement of the print period was the novel and the sophisticated concepts it embodied. (Try explaining to a child the difference between "telling stories" as lies, and "telling stories" as fiction, and see just how complex and abstract the rationale for the novel is.) For two centuries, the novel has been captivating audiences, drawing them in, entertaining and informing them. (Throughout the period when women were denied access to education, the novel was their curriculum, their major source of information and education in values, judgement and character.)

But even the novel has now given way to the television narrative.

Just as Charles Dickens once had a mass audience for his serials, now it is the television series that gets the ratings. Just as readers throughout the country – and beyond – waited with bated breath for the next instalment to find out what happened to some of those larger-than-life, predominantly male, characters (Pip, Nicholas, David, Fagin, Mr Gradgrind, Mr Scrooge, etc.), so international audiences now eagerly await the next episode in a series, to see what happens to larger-than-life stars.

The narrative line of television series and films, however, has not changed all that dramatically from those of novels. (Indeed, a significant number of films has been adapted from novels – and, these days, vice versa.) But the influence of computers and interactive multimedia will transform the "story" out of all recognition in the future.

It is the beginning of the end of the age-old prescription that a story must have a beginning, a middle and an end.

"We are about to enter a world where there is not, necessarily, a beginning, a middle, or an end to anything," says Michael Hiley in his article "Writing for Multimedia"; and "in these uncharted waters of ten thousand tempests, or whatever, the emphasis may well be on the author as navigator, rather than narrator". He continues with his analogy; he sees the reader as "captain, rather than captive".[37]

The big buzz now is not simply multimedia. The new excitement is interactive creations, where the reader – or more accurately, the user-cum-author – interacts with the "story" by means of a menu and a keyboard, or a remote switch. The users can retrieve information in non-linear ways: they can enter into the "story" and take it in any direction that they want to go. They can adapt the old stories – which could mean recycling *Hamlet*, as Shakespeare did when he took the old story and adapted it for the stage. And they can create entirely new stories in new, ongoing, and open forms appropriate to cyberspace.

---

36. Bob Hughes, 1993, "The Parody Machine", *The Electronic Author*, 2, p. 7.
37. Hiley, 1993, "Writing for Multimedia", p. 8.

It means that there won't be a single linear narrative in a singe medium; no story will be set, closed or an enduring reference point. There will be no unchanging wisdom to be transmitted. Rather, "stories" will be fluid and changing, as they often were in oral culture. There will be multiple versions as users make their own marks and create new paths and options, depending on how the mood takes them.

This changes the meaning of a story, which throughout print culture has had the sense of being fixed to some extent. So dramatically does the electronic change the role of writer and reader that it makes these terms almost meaningless in the cyber-context.

Let us start with the multimedia part of the new form. As some commentators have suggested, it is better to think of the multimedia as more like films than books. Talking films, with their combination of sounds, images and print (subtitles and credits) have always been multimedia. Now the electronic book, which can appear on your computer screen from either your own disk or a dial-up database will be able to include *all* the media.

Just as films have relied on teams of scriptwriters, so too will interactive multimedia be able to draw on teams of authors. There are good reasons for this, as Michael Hiley explains:

> We've got to get over the old ways of thinking and working. An "author" won't be able to "write" an "electronic book" just as a film director can't make a film single-handed. An author in the age of interactive multimedia is going to have to know at least enough to commission not only picture researchers but screen designers and programmers.[38]

There's probably no end in sight when it comes to the potential of the new technologies. In talking about the directions of the future, Ross Harley has said that it is the techie's dream to "put everything in the one box – a super-personal-computer-TV-fax-modem-sound-recording-desktop-video-publishing-playback-studio".[39]

This would be the ultimate in empowerment: everyone who had such a box could be a fully multimedia author, able to publish their own productions for everyone else in the world who was wired up.

Not everyone would welcome such developments. This blend of self-created info-tainment is a long way from the pedestal of print authorship, and requires a fundamental cultural shift (for the print generation anyway) before there could be a realistic assessment of its strengths and weaknesses.

But the last thing you would want to say about it is that it means a lowering of standards. That would be exactly the same as arguing in the nineteenth century that there would be a drop in standards if everyone was allowed to read.

To equip today's members of society with the capacity to be authors, film-makers, sound recordists, researchers, graphic artists, presenters, directors, etc., is to have the

---

38. Hiley, "Writing for Multimedia", p. 9.
39. Ross Harley, 1993, "The Nature of New Media Technologies", in Harley (ed.), *New Media Technologies*, Australian Film, Television and Radio School, Sydney, p. 5.

most highly educated and skilled population that could be envisaged. The issue is not that standards are lowered just because everyone can do it; it is that equal access to the technology must be ensured so that no one is excluded from skilled authorship.

The fact that everyone has the potential to be an author doesn't mean that there will be no need for professional authors. Individuals will still play a significant role in developing resources and setting the scene. At least in the next few years, authors will have a crucial contribution to make to interactive multimedia.

Where once there was the passive viewing of television – family audiences spent the entire night transfixed in front of a small screen – the trend over the decades has been towards audience involvement. Even a couch potato can engage in zapping with a remote control, and get the hang of entering and exiting stories at random points by flicking from channel to channel.

The computer screen is another step in the direction of active involvement. I have to keep engaging with mine, or else the screen fades. I am constantly interacting with it, although of course it is primarily a print medium that I am using.

The best current example of interactive media that we have at the moment is the computer/video game. The users are continuously engaged in influencing the action. The characters on the screen can be manipulated to do the user's bidding – within the game's limits. This is the emerging face of the interactive multimedia, the book of the not-so-distant future. Though I suspect today's authors won't find the prospect of being games writers all that alluring.

Gone are the days when children were content to sit quietly in front of a screen. Young people now expect to be involved, to play a part in shaping the outcomes. Print people who take books as the reference point, and regard television viewing or computer game playing as passive, are really stretching the imagination. If ever there was a passive medium in which the audience was led, it has to be that of print. The young are already programmed for multimedia, given all the sound, image, and print sources that are simultaneously pouring into their sensory systems, as they have the TV on, play their music, and do their school assignments – one of the few print tasks they are still required to undertake.

These are tomorrow's readers.

It's easy to see how the passion and possibilities for computer games can become the model for the new "story". The role of the professional author is to help make the game, to coordinate the media, manage the menu, map out the initial possibilities.[40] Then the users of the story make their own choices; they take chances and construct their own version in the process. Then they can send it all round the world if they want to.

Commentators who have tried to describe the new medium have suggested that it resembles theatre more than print. There is an active performance going on, but with one striking difference: the user/reader/viewer/author is one of the actors, as well as

---

40. One very good example of this is *All New Gen*, a game developed by VNS Matrix. See pp. 188–9 for further discussion.

part of the audience, and can be linked in cyberspace to other actors while they are all making the story.

Clearly this new relationship of author and audience will take some getting used to (and will demand a complete rethink of theory, of the traditional separation of actor and audience). There is considerable challenge in the idea of open-ended narratives in which there is no set sequence, but where users can enter the "story" at numerous points, and create their own outcomes.

There will be professional authors, of course. But they won't have much in common with the romanticised model of past centuries. They will be a new combination of artist and scientist, a new breed which has both craft and technological expertise, a new band of infotainers. They will be visually literate, as well as computer and print literate (and "sound-literate" too). They could be the shapers of future culture in the way that authors have been the value-makers of the past.

The new authors will be behind the software and video disks that are the paper and printers' ink of the new media. Michael Hiley is one who believes that the leap into the future is neither too frightening nor too great for contemporary authors to contemplate:

> I don't think authors should worry. I think that the idea of an individual drawing together key elements to produce a finished piece of work will remain of key importance, though I think it is inevitable that authors will have to work in teams of people bringing a great variety of specialisms to bear on the problems of producing a finished piece of work.[41]

Something tells me that most authors won't find this reassuring.

We are in a transitional period, and there are stages on the way to interactive, multi-media authorship. More and more professional writers are beginning to glimpse the potential of the networks, databases and CD-ROMs, and are beginning to play with information technology. Those who have been raised with computer authorship are very much at home with all the new possibilities. Already we have many of the classic stories on CD-ROM, with the options of different perspectives on the plot, different endings, etc. But countless new stories will have to be authored for cyberspace as well.

If we can only speculate on the forms of these new genres, there is nothing speculative about the new forms that technology will provide:

> The computer industry is now hoping that "domestic" computers that look like VCRs and that are driven not by keyboards and mice but from remotes that emulate the CR handsets will open up a massive and lucrative new market as big as or even bigger than VHS. In order to create a demand for such systems, they need an attractive library of publications (just as VHS needed movies on video to get it off the ground). This is where multimedia comes into the picture.[42]

---

41. Hiley, "Writing for Multimedia", p. 9.
42. Paul Brown, 1993, "Digital Technology and Motion Pictures", in Harley (ed.), *New Media Technologies*, p. 93.

Clearly, there will be a great demand for multimedia authors. The question is whether they will be drawn from the ranks of today's professionals, who are primarily print-reared, or whether they will be the graduates of the new media and communications departments. (I am not sure about the status of creative writing courses!)

Some old hands have already made the switch: some want to but aren't quite sure how to go about it: some are hugely resistant. And some are very busy. Most people are surprised to know that there are already more than 10,000 CD-ROM titles available, and the number is about to grow at a staggering rate.[43]

It's not just the authors of fiction who are being challenged by the new multimedia and interactive forms either. The new sources of information (and entertainment) are having profound affects at the other end of the spectrum. Open Learning, for example, has been a huge success and while programs are currently available across the media (with radio, television, print and sometimes oral sources), it is but a matter of time before they are combined. This has had far-reaching consequences for academics who are now sometimes more likely to be found writing scripts than journal articles.

This isn't the only change in their role. As interactive media become more available, even the university lecturers who have provided feedback – in classrooms, on essays, through exams and via the telephone (in distance education) – could find themselves displaced. With instant feedback from the disk, why would students want the teacher's (possibly less reliable and often more arbitrary) response?

So what is the future for academic writing and publishing?

The prediction of the paperless university library has been around for some time, and there has certainly been a huge reduction in library subscriptions to print journals. There has also been a reduction in conventional library funding. At the same time that libraries have cut back their traditional holdings, the resources flowing to computer science and electronic retrieval systems have been more than impressive. It has to be acknowledged that a move to the electronic journal – to electronic storage and access of information in all its forms – is on, and that academic publishing will never again be what it was a decade ago.

The shift has been from "publish or perish" to "perform or perish", and this trend, along with the presentation of PhDs as software, or video, is the substance of the section on electronic scholarship (pp. 120–36). Just as there are changes occurring in the medium, so too are changes taking place in the messages of academia.

Jay David Bolter has some interesting things to say about this. He sees academic writing as having epitomised the mono-dimensional, linear logic of print with the statement of proof, the objective conclusion, the definitive account, the end of the argument and the end of the story.

This particular style will have little place in the cyberworld. Electronic writing, he says (in relation to print), "suggests that the writer need not and perhaps cannot reduce

---

43. Brown, "Digital Technology and Motion Pictures", p. 91.

a complex network of evidence to a single line of argumentation". He urges academics to "extend, modify, and reinvent narrative" so that they can "forge a writing space appropriate to the new technology".[44]

In this new style there is more than one version of the story. The history of Australia, for example, could include the history of white government, of Black invasion; and the experience of women could call into account the official versions provided by men. In this world, there is no one fixed story, and the emphasis is "above all [on] holding things in suspension, deferring conclusions, resisting closure".[45]

If there is any way of representing the structure of the new stories (particularly on paper) it is as intersecting circles, rather than as the linear narrative of print. (Anyone interested in the technical conceptualisation of the way to write for multimedia could think in terms of a tree, with loop branches.)[46]

All the narratives, be they fiction or "factual", imaginative or academic, are going to be changed by being authored in the electronic media. Perhaps professional authors will take up new tools: perhaps it will be the younger generation that shows the way.

The Games Kids Play

MUDs and MOOs and even MUSES are the models for the collaborative and inter-active stories of cyberspace. As pencil and paper were once needed as the tools for writing stories, a personal computer and access to the Internet are all that is required to participate in these text-based games – for hours, and hours, and hours. It's mainly boys who play them; this has its problems.

The MUD – Multi-User-Dungeon – originated in the fantasy role-playing board game, *Dungeons and Dragons*; it was authored in the 1980s by two men (Roy Traubshaw and Richard Bartle) and put on the Internet. It reflects many male interests and has become something of a paradigm; this is also a problem.

Other versions of text-based games have evolved. Developed from *Star Trek* or *The Hobbit* or any of the sci-fi or fantasy novels that are popular, these ongoing stories pull in students from all round the world in their thousands. With the exception of Australian students. In their report on MUDs in the magazine *Wired*, (appropriately entitled "The Dragon Ate My Homework"), Kevin Kelly and Howard Rheingold state that "Australia, linked to the rest of the world by a limited number of precious data lines, has banned all MUDs from its continent: student-constructed virtual worlds were crowding out bank-note updates and personal phone calls."[47]

So how does this virtual-world game work, and why are so many male students hooked on this new text? Why is it that so many of these youths, whose reading and

---

44. Bolter, "Alone and Together in the Electronic Bazaar", p. 11.
45. Bolter, "Alone and Together in the Electronic Bazaar", p. 17.
46. For further discussion, see Bill Seaman, 1993, "Interactive Videodisk Production", in Harley (ed.), *New Media Technologies*, pp. 102–3.
47. Kevin Kelly & Howard Rheingold, 1993, "The Dragon Ate My Homework", *Wired*, July–August, p. 73.

writing skills are often considered below standard by their school, are immersing themselves in a form that calls for sophisticated literary skills, and demands the utmost in concentration and creativity? Why are these players skipping school to spend their time producing, and participating in, such imaginative story-making?

I have no doubt that the allure is much the same as the one which drew me into the fantasy world of books, and which compelled me to produce my own handwritten "novels" as an adolescent. But whereas I took myself off alone and created my own imaginary world to my heart's content, today's youths log on to their computers and travel along the information highways, gathering friends on the way: and together they create their virtual world in cyberspace.

Most MUDs are still text-based. Once you have met your virtual friends, in the virtual place, and given yourself a virtual identity, the game gets underway. Players know they have got to where they want to be because the messages appear on the screen describing the setting, the cast, the script. The screen displays the information, "You are in a long and beautiful tunnel. There are strange murals on the wall." (This is my version and a pure fantasy fantasy, of course, but to my mind it is one that is infinitely preferable to the violent and exploitative descriptions that are typical of male MUD scenarios.)

Someone else has created this space of the tunnel, and the mural (and thousands of other extras and objects). The user can decide what to do in this virtual space. Maybe you key in "Feel mural to see if there is a secret door." The response comes back: "Murals are sacred. Not to be touched." It is up to the user to decide what to do and where to go next.

There are thousands of other players out there, many of them in the same virtual place as you. All of them are interacting, authoring their scripts, setting up their scenarios. But you do not know who all these people contributing to the story *really* are, of course.

In entering the game, I could have given myself the following virtual identity. "Delilah enters. She is tall, dark, clearly athletic and well coordinated. Has thick, abundant hair. Great poise and a gracious manner. Very dispassionate." Such an identity represents a great deal of wishful thinking on my part. I am sure most other characters' descriptions do too. Putting your virtual self into the "story" like this is qualitatively different from being a third person or invisible narrator in print. This is being part of the action, as well as the director of it. Other characters (who could equally well be fantasising about their own race, gender, shape and status) respond to your projected cyber-identity. Like any actor writing their own script, you could come to believe in your created self as you get the feedback to it. The information highways do a much better job than print could ever do in transporting you to another world.

This is not the end of the possibilities. Users can also "pose" or "emote". This is a way of adding a symbolic system that lets others know more than the words could tell, as Kevin Kelly and Howard Rheingold explain.

MUDers use "poses" as well as words to convey meaning and action, giving MUDs an odd but definitely useful kind of disembodied body language. Posing (also known as emoting) . . . adds a new dimension to your communications. Instead of replying to a statement, you can smirk. Instead of leaving the room, you can disappear in a cloud of iridescent bubbles. Emoting seems awkward and artificial at first, but once you get the hang of it, poses give you some added control over the atmosphere in which a conversation takes place – the all-important context that is often missing from words alone.[48]

You can see what the attraction is; why some young people use MUD for hours and hours and hours (when many would argue that they should be outside playing, or doing their homework, or reading a book, or developing their social skills). There are endless possibilities for creating stages, encountering actors, developing dialogue, and exchanging plots. And there's the opportunity for really playing games, of course.

Pranks are also rampant. One demented player devised an invisible "spud" that, when accidentally picked up by another player we'll call Visitor, would remove Visitor's limbs. As this happened, others in the room would read "Visitor rolls about on the floor, twitching excitedly." Worried players could summon a wizard or god to fix Visitor, but as soon as they "looked" at him, they too would be spudded, so that everyone would then read "Wizard rolls about on the floor twitching excitedly."

Ordinary MUD objects can be booby-trapped to do almost anything. A favorite pastime is to manufacture an object and get others to examine or use it without knowing its true powers. For example, when you innocently inspect a "Home Sweet Home" cross-stitch hanging on someone's wall, it might instantly and forcibly teleport you back "home" to the beginning of the game, while flashing "There is no place like home."[49]

MUDs have been played mainly by young men – hence the gore, and grotesque and gratuitous violence. But a new wave is making its appearance with much more artistic and scientific goals.[50] Named MOOs (MUD Object Oriented), and Muses and Jupiter (an image database system for astronomers) the new games and facilities are serving a range of experimental communities and virtual societies (At a slight stretch of the imagination, they could be said to parallel some of the sub-genres within the novel, and the wider print repertoires.)

For example, there is an educational cyberspace called Cyberion City, where the stories that have been created by kids are nothing short of extraordinary.

On any random day 500 kids beam up to Cyberion City to roam or build without pause. So far the kids have built more than 50,000 objects, characters and rooms. There's a mall, complete with multiplex cinema (and text movies written by kids); a city hall; a science museum; a *Wizard of Oz* theme park; a CB radio network; acres of suburban housing, and a tour bus. A robot real estate agent roams around making deals with anyone who wants to buy a house.[51]

---

48. Kelly & Rheingold, "The Dragon Ate My Homework", pp. 70–1.
49. Kelly & Rheingold, "The Dragon Ate My Homework", pp. 72–3.
50. See Amy Bruckman, 1993, "A Study in MUDs", *Wired*, July–August, pp. 70–1.
51. Kelly & Rheingold, "The Dragon Ate My Homework", p. 70.

The complex and interactive story-lines which have created an entire community are the work of young people who, not so long ago, would have put the same energy and skill into writing their "novels" to share with other members of the class. I know that where once I wrote for hours and hours in my exercise book, constructing characters that were real to me, and developing plots which fascinated me (and no doubt met some of my psychological needs), these days I would be spending my time on the Internet. I am sure that I would find it as much of a mind-blowing leap as those early readers did when they went from manuscripts to books.

Nothing that went into my written stories would have to be abandoned if I were to take up screen storying; the interaction, the feedback, the contributions of others would add so much to the experience. Even now I am tempted to become involved when I see something like the following prompt on my screen. I know that this is the educational publishing/writing of the future.

Welcome to MicroMUSE.: we are hosted at chezmoto.ai.mit.edu.port 4201.

* *

REMINDER: Read 'news' regularly to keep up on changes and additions to the server. New commands will be listed in 'news' with details provided in 'help'. For more information new players should type 'help'

getting started

* *

Cyberion City Main Transporter Receiving Station

The bright lights of the Cyberion City Transporter Station slowly come into focus. You have been beamed up here (at considerable expense) from one of the Earth Transporter Stations. You are among the adventurous and moderately wealthy few who have decided to visit (and perhaps dwell) in Cyberion City, the largest space city in the solar system. You are welcomed by the transporter attendant, who gives directions to all newcomers in this space city.

Contents: Attendants
Obvious exits: Out

Welcome to MicroMUSE. Your name is Guest.

Attendant says: "Welcome, Guest, to Cyberion City."
Attendant says: "Feel free to contact any official for aid."
Attendant says: "Be sure to use our extensive on-line help command."
Attendant says: "I hope you enjoy your stay."
The attendant smiles at you.
You step down off the MTRS platform.

Main Transporter Lobby

This room has high, vaulted ceilings and white walls. The thick, black carpet makes no sound beneath your feet. You are just inside the transporter lobby, where Visitors arrive from Earth. To one side is an Information Desk. A door leads to the Tours Office, and another leads Out into Cyberion City proper. A Public Relations Dept. Intercom stands in the center of the floor; type "look Intercom" for instructions.[52]

---

52. Kelly & Rheingold, "The Dragon Ate My Homework", p. 71.

No educator, publisher, or author, can ignore the magic and the pull of such a medium; and this is the real-life experience of many of the younger generation at this moment. (Assuming they have the resources, of course; but it takes only a computer, a modem and an Internet account – as readily available as telephone accounts in the US – and to type in an address, to participate in this creative process.) Already, Cyberion City has become the place where enlightened parents send their children, rather than to conventional school. Run by MediaLab at the Massachusetts Institute of Technology, Cyberion City provides a bedazzling positive resource; it has a charter which all guests and users have to read, and which (for obvious reasons) bans violence and killings.

If there is any electronic model of authorship, this is it.

The fact that everyone will be able to play does not pre-empt the development of professionals; authors who create breakthroughs, who test the medium to the limits, who bring their own fantasies and skills into the process. The next generation of writers are serving their apprenticeship now, not by handwriting novels in exercise books but by beaming up to Cyberion City.

The kids are playing games in cyberspace, and they can teach the print generation a thing or two about authorship, along with many other skills.

5

EDUCATION

From Factory Model to Cyberspace

Travelling to the Experts

As the description of Cyberion City (Chapter 4) indicates, if not today, then in a very short space of time, students won't need to go to school or college for information.

We might be getting used to the idea that educational institutions don't need a physical place called a library (all they need is electronic access to a library, or database): but how open are we to the idea that we don't need a physical place called a school, a college or a university? That all that will be required is for students to have access to databases?

To some extent, we are seeing this happen already as students take up Open Learning options in Australia, and with the aid of television, radio, print, and e-mail (and sometimes face-to-face contact) they independently undertake their own learning. But these are early days: fairly simple programs are being offered, and they are still ordered and controlled by state systems and educational institutions.

Nonetheless, they often feature some of the best possible teachers, who attract a range of listeners and viewers. The audience for these publicly broadcast lessons is by no means confined to the officially enrolled students. Many members of the community find the dynamic professional presenters, the articulate and telegenic experts, and the topics from science to sociology, good *entertainment*. They enjoy what is being presented. And this helps to give an idea of the role and impact that on-line databases, interactive disks and multimedia will have on future educational arrangements.

Why would kids want to go to a boring place called school, where dumb old teachers insist on regimentation, and try to make you be quiet, read books, use paper and pen to copy notes from the blackboard, and where you are forced to do – handwriting! – when you can be connected to your friends in front of a screen, or networked at school, and access some of the most exciting information in the world? When you can be "infotained" to your heart's content, and with the aid of a remote switch, control all the pathways, the "stories", the games that you find thrilling? When you can be in charge of what you do!

As an adult student, why would you want to travel all the way to a university campus, where you can't park: to sit in a stuffy lecture theatre and listen to a droning monologue on politics, delivered in a deadening monotone, which reveals that this is the *n*th year that the same old information has been given out? Where you are supposed to take notes, to "learn" later?

Why would you go through this when you could dial up a database or get on the Internet, and be "edutained" by some of the most stimulating political personalities? You could have Glenda Jackson on the history of democracy, Dr Carmen Lawrence on psychology of the political scene, or Barry Jones on the art of the right question. And all presented with the panache and professionalism of the latest film techniques. On-line, multimedia format – *and interactive*. Which is more than can be said of the average lesson or lecture!

The alternative already exists. You could have a CD-ROM which lets you learn (or engage with, or use, or research) all the information[1] you want. In your own time. From a much more exhilarating and enjoyable package of much better quality than has ever been provided in a school or college.

So why would you physically travel to an educational institution when you can do some virtual travel to cyber-education? Why would you listen to teachers in class-rooms? Why would you want to *learn* the limited curriculum that is currently imposed in regimented educational institutions?

This is not to claim that there won't be schools or colleges or universities in the foreseeable future, although they are under threat, and they might look unrecognisable.[2] For many reasons (some of them less than admirable) we will probably continue to want young people "institutionalised". But this is to suggest that educational institutions are going to have to find new reasons for their existence and new functions to fulfil. Schools and colleges and universities are going to have to undergo radical restructuring right now, or else find themselves without a role. Because if they are not places to go for information – and for the skills on how to get it, organise it and utilise it – then what are they there for?

The Need to Know

Classrooms were established primarily for the purpose of passing on information. Virtually every educational theory and practice has been based on the premise that there is a *body* of knowledge which the teacher has mastered, generally from books, and which the student is required to "know". We have a model in which teachers hold the information in their heads, and teaching consists of transmitting it to the students. They, in turn, are obliged to store it in their heads, and then regularly recall it when

---

1. *Information* is being used here as a generic term; it has no connotations of good or bad – it is just data.
2. For further discussion on the unrecognisable nature of universities, see pp. 137–46.

they are tested. This all takes place in institutions.

It's hardly a model that has appeal for the electronic technology generation.

"Historically, education has been a passive experience for most students," states Margaret Fallshaw, computer consultant at a school that has adopted the new medium. "Most teaching continues to be offered in classrooms, with groups of learners being taught by an individual teacher. This age-old practice is probably the single most effective deterrent to improvements in learning."[3]

That's not the end of it, either. It's not good enough for kids any more. It's a teaching/ learning model that is out of synch with the rest of their world. Many of today's students can tell you in no uncertain terms just how "unreal" (and boring, and silly) the educational context is. Traditional educational theory, practice and organisation are each day becoming more irrelevant and unworkable: just as the scribal model became obsolete after print was invented.

We have moved so far into the electronic revolution that it is almost with the benefit of hindsight that we can – if we want to – "look back" on the way education has been conducted for the last century or more. (If there is any one origin of the system we have inherited, it is that devised by Matthew Arnold in the 1840s for Rugby, one of the private boys' schools of England. It was intended to train "rulers" for the Empire.)

Even a cursory analysis reveals the enormous inadequacies of the present system in the information world. We are confronted by a growing need to restructure the educational establishment and to rethink the role that we have traditionally set out for teachers – and for students.

At the centre of all our received wisdom about teaching and learning are some very simple premises that are consistent with the standardised cultural mindset of print. This is to be expected.

One of these premises is that knowledge is a *body*. We assume that knowledge exists as an entity: that it is stable enough to be divided into disciplines; that it can be arranged in hierarchical order, and systematically taught to various grades in 50-minute segments, on certain days of the year. But this conceptualisation of knowledge is increasingly inadequate in an electronic age in which knowledge is also information which flows in cyberspace; it can be public, it resists boundaries, it is almost impossible to rank, and it is available twenty-four hours a day.

In an attempt to identify some of the differences between the old and the new information – and the way people "use" it – Peter Lyman (University of Southern California) did a study with the educational publisher, McGraw Hill. The idea was to see if people used standard textbooks differently when they were presented on a screen and could be accessed electronically. The results were quite shocking to some of the traditionalists.

---

3. Margaret Fallshaw, 1993, "The Promises of Educational Technology", in Irene Grasso & Margaret Fallshaw (eds), *Reflections of a Learning Community: Views on the Introduction of Laptops at MLC*, Methodist Ladies' College, Melbourne, p. 17.

Publishers and teachers assumed that the users would move from Chapter One, through to Chapter Two and continue. As books lead you to do. But the students darted all over the place, making full use of the electronic capacity; they went from paragraph to paragraph, "taking everything out of context".

> This horrified the McGraw Hill writers, because they had thought they were writing in a narrative form in which knowledge would be cumulative. . . . But users treated narrative argument as if it were a database, selecting paragraphs without context . . . In the on-line text, the reader designs custom information, selecting it, editing it. [4]

Several of the McGraw Hill writers withdrew their work because they were not willing to have it taken out of context; they wanted to be the authors, and to have their material stay in the shape that they created it. They didn't want the students "messing" with it. It was different information when it was read out of order, and added to, synthesised and changed.

As Peter Lyman says, the way in which students use information on-line makes it "a whole new ballgame". "We can no longer describe [students] as readers because in an important sense they are authors," he comments. This raises a serious issue, not just for conventional authors or publishers but for all educational practitioners. "Higher education and the scholarly disciplines are organised around the concept of a literature, which is to say an authoritative, cumulative version of knowledge; but that is not the way people are using knowledge on-line."[5]

Another premise is that teachers are the ones who know, that their task is to channel what they know into the students. Students are supposed to sit there passively in serried rows and take all this in. They become knowers when they store all that they have been taught in their own heads.

This theory about knowledge and its transmission is also becoming increasingly difficult to sustain. In the electronic world, we don't have knowers: we have *users*. The transition has been going on for the past few decades: the role of the teacher as authority figure and knower has been in decline as the amount of available information expands. As most teachers are print-reared, and as only a relative few have made the shift to the new media which now is the business of the world, the gap between what teachers know and what is electronically produced is forever widening.

Even if teachers were to make the shift and become users, and part of the new culture, they would still not be able to regain their status as knowers, as people in possession of a stock of approved information to pass on. Apart from the fact that "someone who knows" is an irrelevant category in a computer-based world, there is the issue that it is simply not humanly possible to be a knower any more. There is just

---

4. See Peter Lyman, Follett Lecture, Internet, 4 October 1994, directed courtesy of Janine Schmidt and Colin Steele.
5. See Lyman, Follett Lecture.

too much information for any one individual to master. An average ten-year-old, addicted to a database on dinosaurs or dragons, on rocks, rainforests or religion, could expect to be better informed than the teacher – particularly if the teacher's source of information is print. When you multiply this by every student in the class – each doing their own electronic thing – it simply is not possible for teachers to appear as experts, to be able to "correct" the student, and pass judgement on "right and wrong" in relation to content.

Within the print-based education system, knowledge was considered something special. It was in short supply and you had to prove your fitness to be entrusted with it. If you wanted access to the next level of information, then as a student you had to pass a test to qualify; those who didn't were not allowed to continue. Only the few could go on to university; only the chosen could become postgraduate students. This is dramatically different from today, when there is so much information around and so few entry requirements for those who want access to it. For example, there is no test for TV viewing.

This change has led Peter Lyman to talk about the "ecology of knowledge": while knowledge was scarce, it required centralisation and conservation, he comments. But now that we live in a data-saturated environment, "our distinctive information problem is one of selection, as much as conservation".[6] Very few of our educational practitioners are in a position to appreciate the significance of the changes that have occurred. They continue to have certain ideas about knowledge and standards and teaching and learning that have little relevance in the electronic world, and that serve to distance them even further from their students.

The graded curriculum where students are to study a specific period or problem one year, and move on to another the next, looks increasingly absurd as kids dial up databases on whatever takes their interest, and become independent learners (or users!). Even TV has dramatically altered the nature of school knowledge and made nonsense of its arrangement. What took me years to learn in biology, in senior high school, was covered by David Attenborough in two hours of television, in ways that were more informative, more enjoyable, and much more meaningful and memorable.

And this contribution of the electronic media has completely eroded the knowledge-control of educational institutions. You don't have to wait now until Grade 10 to get the lessons provided by David Attenborough, or those presented in "Quantum", or "Images of Australia', etc. on Open Learning. Even young kids can make sense of many of these programs, and this makes a mockery of the age-grading system. It destroys the credibility of the education enterprise. It challenges the educational hold on knowledge. *Competency*, not qualifications, becomes the only criterion.

School authorities who persist with the old model and try to retain their monopoly on information, who dole it out in their own time and on their own terms, are putting

---

6. See Lyman, Follett Lecture.

themselves at risk. Schools where staff insist that students wait until it is considered the right time to make certain information available to them are in danger of becoming a laughing-stock as well as irrelevant. To tell students that something they have seen on television (or encountered in a game) and about which they are curious isn't covered until they reach a grade a few years hence, is to lose their interest and respect. And because of the requirements of some systems, this is the current climate of too many classrooms.

This isn't the only area where traditional educational knowledge has fallen into disrepute as we become surrounded by electronic information. There are countless examples where the curriculum has had to be changed, and where examination questions have had to be revised, because kids have been learning from electronic sources rather than restricting themselves to what is provided in educational institutions. (This is another version of the school not wanting parents to teach their kids to read.)

There's no point in setting an exam, when everyone knows the answer because it has been on television. If everyone passed, the examination would be of no use. It wouldn't tell you who should go on, and who should be eliminated. So, in the attempt to keep ahead of the media – and to maintain their standards – educational organisations are often engaged in an undignified scramble to find testable information that isn't freely available on the air waves. In itself, this is a long way from the publicly stated aims and purposes of exams and education.

We are only just beginning to glimpse the way cyberspace will bring an end to content-based examinations. What can be stated with certainty is that we are about to witness a challenge to the most deeply entrenched values of the print education system – and it is not likely to be the students who fail the test.

The Trouble with Heads

The teaching/learning model which currently predominates is one in which learning is held to take place when students have memorised, stored, and kept in their heads, the knowledge that their teachers have taught them. This is why students are regularly required to recall this information: and they are tested, and credentialled, on their ability to do so. Frequently, they can be judged to have passed or failed, depending on what they have retained in their heads, and the extent to which they can accurately represent it.

Such a method is a passive (and partial) model of learning. It is one which David Loader (the Principal of Methodist Ladies' College Melbourne, a school which has embraced the computer era) derisively describes as an "industrial model". Too often we have a factory model of education, he argues: the emphasis is on "'batch processing' of groups of students in classes"; "the measurement of a teacher's workload [is in terms of] the number of classes taught"[7] and the measurement of student learning is

---

7. David Loader, 1993, "Reconstructing an Australian School", in Grasso & Fallshaw (eds), *Reflections of a Learning Community*, p. 10.

the number of examinations passed. Where the world is divided into right and wrong. Every aspect of this model is now under threat.

In an electronic era, an education system which is based on what people keep in their heads is doomed. If there is one thing that computers have taught us, it is that heads are not good places for keeping information in. In comparison to electronic retrieval systems, heads are poor storage systems.

For the last few hundred years the "body of knowledge" idea that we have assumed has led us to believe that knowledge is fixed, that it can be stored in books, and can be transferred to heads. Of course we have always had some practical subjects (chemistry) and some hands-on professions (engineering, dentistry), where we have encouraged students to be *users* of information. But this is permitted only after they have reached a certain stage of knowing what's in the books. Within the book-based culture, they have been receivers of information – not creators of it.

This sort of knowing isn't going to be of much use any more. No one can know enough in cyberspace. No matter how extensive their learning, and how good their memory, no one will be able to compete with the computer for recall.

In this process, the very nature of human beings as we have understood them is being transformed. Our beliefs about knowledge, about teachers, students, and the way learning takes place are all undergoing drastic restructuring.

It's not much different from the changes which took place during the earlier information revolution when print was introduced. The monks who had made such a ritual of reading, and who insisted that scholarship came from knowing the sacred texts – off by heart! – were appalled by the changes that confronted them. The prospect of students going off to read on their own, racing through one text after another, filled them with no less horror and despair than that experienced by the McGraw Hill authors. (Readers who have chosen not to progress through this text in a linear fashion will find this example on pp. 101–2.)

These new book readers, the monks insisted, couldn't possibly know what they were reading. They went through too many texts too quickly. They would never be able to memorise and recite "the word" they had read. And to the monks this was the end of everything they held dear. They could only see that what was going on around them as a drop in standards; the work of the devil; the abandonment of learning and scholarship, of all that was the highest and the best in human beings.

What we can see now is that with the new medium, the knowledge of the monks no longer counted. Memorising was no longer an appropriate activity once the comparatively few manuscripts gave way to the thousands (millions) of secular books. Reciting out loud to fellow scholars was no longer a valid test of understanding and learning, once everyone could go off and do their own thing and form their own conclusions. An entirely new education system – state schools and universities – evolved outside the control of the Church, where the new notions of information and knowledge, and new methods of teaching, inquiring and learning could flourish.

Such a fundamental educational change is taking place again. This is because teaching knowledge solely from books is no longer appropriate now that there is the electronic medium. Teaching students to store information in their heads – and to recall it on demand, on certain days of the year, with pen and paper (which is still the way that education is conducted in many places), and then to label them correct or incorrect, is neither a valid nor a useful activity in the computer era.

If we are perceptive we can see that a new, open, electronic system – outside much of the control of the traditional authorities – is beginning to emerge. And it is set to expand dramatically.

When young people today give every sign that they would rather "do" than sit and "learn", this is not a drop in standards that we are witnessing. It's not a sign that they can't concentrate. (Nor do they they have Attention Deficit Disorder, for which they need drug treatment.) It's not a sign that the next generation is backbone-deficient, nor does it mark the end of civilisation. It's not the work of the devil that makes students prefer to keep information (from spellcheckers to multiplication tables), in their machines rather than in their heads.

What it does mean is that we have to reassess our educational system and our theories of learning and teaching. We have to recognise that the essence of human history is the creation of more and more forms of communication – and the retention of less and less in our heads. In oral cultures, for example, there is complete dependence on memory: this is where "the survival of knowledge depends upon the survival, and its dissemination upon the mobility, of its human characters".[8] In technological cultures, very little of the vast amount of information created depends upon memorisation. Yet every change along the way has met with protests and predictions of the end of human communication.

Socrates was against writing because it reduced the role of memory. The monks were against the book for the same reason. And some people today are against the computer because it further reduces the importance of memorisation. All this tells us that there are those who have amnesia, who don't know their history, and who aren't prepared for change.

What we have to do now, as a result of the latest changes, is rethink the relationship between human beings and machines when so many members of society are "wired up" or "plugged in".[9] And these issues can be at the heart of new subjects that will emerge in the computer-based education system.

Books for Knowing: Computers for Doing

Many young people already operate in the new medium. (They have not had to make a transition: it's the way their world works.) They understand that information is free-

---

8. Judy Marcure, 1993, "Dynamic Documents and Print Conventions: New Directions in Scholarly Communications"; private communication.
9. See for example, Donna Haraway, 1985, "Manifesto for Cyborgs", *Socialist Review*, 80, pp. 65–108; and Marge Piercy's novel, published as *He, She, and It* in the USA and as *Body of Glass* in UK and Australia (Penguin, 1992).

flowing (in the ether), and that the issue is to access it. They don't have the same commitment to keeping things in their heads, regardless of whether they are working on a computer, cruising on the Internet, watching television, playing video games or reading a book.

It's the games that provide the clue to the human changes that are in progress.

There's no substantive "information" or "knowledge" in them; nothing to learn off by heart in the traditional sense. *It's what you* do *in the game that is important.*

It's not a matter of right or wrong but how far you can get, how long you are involved. What counts is the skills you develop, the judgements you make, how you respond and interact. This is why kids are puzzled when their elders ask them what they "learn" from games; it is also why they are often antagonised by the dismissal of games as mindless, a waste of time, and a pursuit without skill. Any adult who assumes kids prefer games to books because they are "easier" or because they are mentally a "cop-out" indicates just how little they understand about the electronic revolution.

For at this stage kids know something that their print-preoccupied elders often do not: namely that games are intellectually demanding; that they make you think, assess, react, decide, *and act* in a way that is rarely the case in an educational institution. Educational theorists and policy-makers may be shocked and affronted to the core at the very idea that the insights for teaching/learning activities of the future will come from studying kids and their computer and video games, but there's no doubt that this is where the data are to be found.

The evidence is already there for those who choose to look. What we know and how we know it is being transformed as we move from a print-based to an electronic society. Whereas books have been for knowing – the emphasis has been on knowing what's in them, being able to quote from them and relate the contents to an audience – this is not the way the electronic media work. Books are for knowing, but computers are for doing.

You don't need to know when you can constantly look something up, when you can play it again; you don't need to be able to recall when you can always retrieve.

Some organisations in the community are already aware of the shifts that are under way. The advertising agency McCann Erickson, for example, has done the research which puts it streets ahead of most education departments when it comes to the way kids are processing information. In its Australian study "Out of the Real", research director Dave McCaughan found that for 16- and 17-year-olds, video games were the basic level of technology and that the students were video literate. They "preferred information to be imparted as video", he explains, because "it is easy to understand after years of playing video games and watching television".[10]

While educators lament the drop in standards, and try even harder to get kids to read, to "digest", to know and recall, advertising agencies are shrewdly assessing the most palatable ways of presenting information to the younger generation. Instead of

---

10. See Monique Farmer, 1993, "PCs Can Be Cool", *Sydney Morning Herald*, 8 February.

sticking with the received wisdom of the past, they are starting with the kids themselves to see what attracts, and what works.

According to Dave McCaughan, the 16- and 17-year-olds are of one accord: "If something lacks video action and interactivity, it's boring. Some of the comments suggested that any technology involving script, such as text on a computer screen, was old fashioned and *redundant*" (my emphasis – the kids don't think they need print!). You can bet that McCann Erickson isn't going to advise its clients to use print to reach their markets!

"Today's teenagers approach computers and electronic equipment with enthusiasm, not fear, and are so confident they rarely consult manuals," Dave McCaughan reports. And he shows how the young have moved from "knowing" to "doing":

> Most don't even acknowledge that there are manuals. They seem to think what's the point of having any electronic thing if you've got to read a piece of paper to use it? Generally they're quite willing to just jump in and start playing around."[11]

(He didn't say whether he was referring to girls or boys, however.)

Having watched four-year-olds at the computer who solemnly informed me that "you can do a lot with a double click", I have no trouble agreeing with Dave McCaughan's findings.

Young people are *doing* information and learning. They want to have information literally at their fingertips, and to be able to work it, move it around, play with it, rather than keep it in their heads in static form, and deliver it on cue. They find that information processing can be exciting and entertaining, in contrast to print which they can experience as boring. This has revolutionary implications for education.

Think what this new form of learning/using is going to do the content-based curriculum and to examinations. We have already witnessed quite a kerfuffle about calculators, which was only the start of the problem. First, it was argued that students would become mentally lazy (and develop slothful habits) if allowed to rely upon a calculator. Then they would be deficient: for without information in their heads, how would they be able to calculate *if there was no calculator present!*

This is like being against teaching kids to read on the grounds that it would make them book-dependent.

But the most heated arguments of all have been about allowing kids to take calculators into exams. What sort? How powerful? Does every student have one with the same capacity? How can you check? And how will the exam questions have to be changed if students can use calculators?

Again, there would be no point in having exams if every candidate could use a calculator to get the right answer. There's no point to the whole procedure unless some students get it wrong. In educational circles, if all candidates got 100 per cent it would

---

11. Farmer, "PCs Can Be Cool".

not be seen as a great achievement, or as an increase in skill level throughout the population because everyone could now do what only the privileged few could do but a couple of years ago. It would be seen as a useless exam. Because in the education system of the print period, the reason for exams has been to sort those who pass from those who fail.

This could be a bit difficult to organise with computers.

And I bet it isn't long before we have some print-reared educators who seriously declare that even the dullest students can manage to solve problems once they have the right software; so, sadly, you can't tell who the "bright" ones are any more.

I am sure that there are educators who will never give up their precious exams, not even long after it has been demonstrated that they don't test what they are supposed to and that they simply aren't workable. I can envisage speed tests, for example, as students arrive in examination halls with their Laptops and their own programs, and disks, even modems. The test will be to see how long it takes students to retrieve a specific piece of information or to work out a designated problem on their computer. Perhaps this will be the interim measure as we move from content-based exams to the daily activity of doing research, learning, thinking, using, whatever you want to call the new form of intellectual employment. But no matter how you label it, it's a method which doesn't lead to "correct" answers or standardised responses.

Despite the education system's history of being able to reduce almost everything to testable proportions, I genuinely believe that this tradition will be thwarted by the computer. We won't be able to have exams and scores in the way that we have grown used to while immovable print has been the medium.

With interactive educational materials, there won't be a ready place for testing to be fitted in. The interactive process has feedback built into it – that's what it is; so the overriding concern with right and wrong is giving way to the principle of whether something works. For the student at the computer with interactive multimedia capacity – who is doing anything from the discovery of DNA to how to build a wolf-proof house having just listened to the story of "The Three Little Pigs" – there can be no conventional test. There is only the excitement of doing, the satisfaction of solving problems, and the continuous feedback that keeps the process going until you decide to quit.

Of course, such competency could become the exam. Students could be given a task – a problem to solve – and access to a computer, with "supervisors" in the room. But this would mean for the students that there was no dividing line between the learning and the examining. There would be no distinction between the education and the entertainment. They become all one and the same thing. This is an absolutely revolutionary concept in most parts of our current education system.

It would also mean that more boundaries were being broken, and that educational research and theory which have systematically separated the knowing and the testing would have fewer opportunities for playing such a role. (It could, however, contribute to understandings about competence and credentialling; but these are very different areas.)

There are a few signs that enlightened educators are appreciative of these changes going on under their noses. As is always the case, there are visionary and inspiring teachers – and a few leading schools – where computers are being taken up enthusiastically and critically, and where staff and students are entering the computer age together. But sometimes this is in spite of the existing system. As one director of a teacher training program put it, "We are not doing anything differently from what we did twenty years ago. And it is a national educational disgrace."

Some state systems – in Australia and overseas – are even presenting a "back to basics" approach (which looks like the nineteenth-century model) as a counter to electronic change and what they regard as the "drop in standards".

There are few institutions which have adopted the computer as a medium (which is very different from using it as a tool – to present traditional assignments more professionally, for example). In these institutions, a different educational politics is beginning to emerge. It is one in which teaching models of the past are referred to as *instructionist* – where teachers instructed and the assumption was that students learned what teachers taught. In contrast, the new model of learning is called *constructivist*; constructivism as David Loader explains, means "built by the learner, not supplied by the teacher".[12] This is indeed revolution.

Schools

On Ladies and Laptops

There are of course enormous problems associated with moving from a print-based education system to an electronic one – the most dramatic one being what to do with teachers. But as the case study of Methodist Ladies' College Melbourne (below) makes clear, the terrible consequences that have been predicted as part of the computer scene have not happened. Instead, there is enthusiasm from staff, students and parents about the world which has been opened up to the girls: and about the intellectual and creative growth that the computer-based education has fostered.[13]

In 1989 a Year 7 Class at MLC were provided with their own laptop computers. Laptops, because of space considerations; and their own, because it was believed that every girl should have her personal computer in the way that students have had their own notebooks in the past. David Loader, the principal of the school, explains that it is a *personal* computer, "in that it allows its user to create a personal 'knowledge space' with idiosyncratic ideas, data and software of personal value. The ownership involved

---

12. Loader, "Reconstructing an Australian School", p. 11.
13. The cooperation of the staff has been actively sought, and while there are still some who see the computer as an "extra" rather than the medium, and a few who are convinced that the new is not as good as the old, it can be seen that it is possible to change the culture and the teaching practices, in a relatively short period.

is not just of a machine, but of knowledge and power."[14]

(Having a personal computer, organised the way you want it, is very different from using a school computer that is programmed for use by everybody.)

The purpose was never to teach how to use computers in the way that schools once taught students how to use typewriters. It was never to "create an artificial range of good computer users – kids who will earn an 'A' in computing and others who will earn an 'F' in computing," states educational consultant on the project, Gary Stager.[15] The purpose was to have students *do* learning for themselves, with the aid of a computer. With this in mind, the school purchased licences to use Logo Writer and Forte, and databases (Crossword Maker, The Children's Writing and Publishing Centre, etc.) so that the girls could start making – "constructing" – their own learning.

And the verdict?

Electronic Classroom

The response has been positive on every count. The consensus is that the students are more sociable, that there is more classroom interaction as the girls become more involved, get enthusiastic, share ideas. It's obvious that individuals proceed at their own pace, and in the words of Pam Dettman, the head of the Junior School at MLC,

> we have also noticed that students' problem-solving abilities have improved, because they are generally more prepared to persist with problems on the computer in order to find a solution. Again, we basically attribute this to the immediate and non judgmental feedback, combined with students' non-stop access to the computer.[16]

What about fears that computers will make students mentally lazy, and teach them bad habits; that they are a distraction, a game, rather than the real thing which is study associated with a book? What about the concerns that the machine will put an end to imagination, originality, creativity?

> Creativity is another area that has developed. At first many parents and some staff were concerned that students' creativity might be adversely affected by computer use, but now we feel that the opposite is the case, partly because of the increase in risk-taking, partly because simulations allow students to vicariously experience many things they could not otherwise experience, and partly because the ready use of graphics aids the imagination.[17]

One has only to observe a class of students doing schoolwork with the same commitment, concentration, enthusiasm and sheer enjoyment that we normally associate

---

14. David Loader, 1993, "Resourcing the Future", in Grasso & Fallshaw (eds), *Reflections of a Learning Community*, p. 32.
15. Gary Stager, 1993, "Computers for Kids – Not Schools", in Grasso & Fallshaw (eds), *Reflections of a Learning Community*, p. 39.
16. Pam Dettman, 1993, "A Laptop Revolution", in Grasso & Fallshaw (eds), *Reflections of a Learning Community*, pp. 28–9.
17. Dettman, "A Laptop Revolution", p. 29.

with video games, to know that this is one of the most intellectually expanding and demanding of processes. (An education system which has traditionally functioned on the premise that anything that is fun cannot be proper learning – that it has to "hurt" to be good for you – will have difficulty adjusting to this new style of ebullient intellectual activity.)

Some educationists who have observed computer-based "constructivism" are as excited about it as their students: they recognise that what they are seeing is genuine student involvement with problems; where every ounce of concentration, ingenuity, lateral thinking and judgement is going to make connections, to work out ways of dealing with the issue. "We are watching students learn to think – probably for the first time," comment teachers who are part of the new era.

MLC Melbourne is a school which has gone over to computers. The computer is the medium and students are empowered to set off in their own directions with teachers as support. There are no lessons on how to use a mouse: no forty-minute periods on the "correct" way to solve problems. Instead, packages such as Logo Writer have been made available to students to do their own learning; consultant Gary Stager explains:

> Logo Writer is a popular software package that combines the power of the Logo program-ming language with word processing, graphics, animation, and music in one user-friendly environment. Five-year-olds and university professors experience the same playful enthusi-asm towards problem solving and learning when working with Logo Writer. The learner is free to express himself [sic!] in unlimited ways – not bound by the limits of curriculum or artificial school boundaries.[18]

His claims are not exaggerated. It's not possible to watch young people (or old people!) working with computers on such projects without being impressed by their involvement and sheer delight in what they are doing. "A student working on a traffic management problem using Logo Writer was heard to say in a moment of great excite-ment 'You beautiful little turtle. My God, I love this turtle.'"[19] It's not like learning – and classrooms – used to be. And it's not just the students who are changing. The whole basis of school administration and classroom management has been transformed

On my visit to MLC Melbourne, one of the first things that struck me was the absence of desks. There was nice carpet in the room, and many of the girls who were engrossed in their laptops were lying on it. As a well-trained teacher of the print era, my first reaction was – shock, horror! Why were they not sitting at desks? What could they expect to learn when they were sprawled all over the floor? Apart from the fact that I was soon laughing at my own deeply entrenched responses, I learnt a valuable lesson. When I asked one student what made the floor so attractive, she looked at me in astonishment and said:

---

18. Stager, "Computers for Kids – Not Schools", p. 41.
19. Loader, "Reconstructing an Australian School", p. 11.

But if you sit up tight and straight at a desk and work at a laptop all day, you soon get a sore back or neck or arms or hands. It doesn't happen though when you can move around, and relax and stretch. It would be just plain stupid to sit at desks for hours and hours. The floor – and with your friends – is much better.

Classrooms don't even look like classrooms any more. And, having got over my initial reflex action, I have no regrets.

Removing the Boundaries

But it is not just the desks that have gone. One of the other things to go very quickly, is the 40–50-minute period of instruction. Cheryl Weiner, project director of Multimedia (being developed by Paramount in the USA), says, "One of the biggest problems [with electronic media] is the 40–minute hour . . . You sit down, you get started, and the bell rings. Just how realistic is that?" she asks. "Classrooms with walls and time periods are the main obstacles to using technology well."[20]

At MLC, the school has already moved to a more flexible day. (You don't want students picking up their laptops at the sound of a bell, in order to move to another classroom to work on the same problem. A better arrangement is to get rid of the boundaries: have students stay in the one space with the teachers on the move, and no bells ringing. Given that subject divisions are evaporating as the computer becomes the medium for education, it could even be desirable to have a number of teachers of traditional subjects available in the one space for support.)

The girls have stopped thinking in terms of subject periods. "The timetable has been modified to allow for more sustained work . . . The result is that students will often work into their lunch-time because their work interests and challenges them."[21] And the same can often be said for the staff as well. It's like an educational dream come true!

School time and space aren't the only casualties of the laptop era: school subject division is crumbling. The world won't conveniently stay divided into history, geography, or even mathematics and French, as the staff discovered once the MLC girls took over the responsibility for their own research.

David Loader explains how things "grow"; some of the teachers thought that the Year 8 students could benefit from a French version of Logo Writer, so that they could talk to their laptops in another language. "However, the surprise was that the students were 'talking to each other in French' in maths and science classes. What is more, the maths and science teachers began trying out their French."[22] And the students started doing their maths and science in French. The "veteran French teacher later reported that she had never [before] witnessed students of this age actually speaking French outside a French class . . ."[23]

---

20. See Domenic Stansberry, 1993, "Taking the Plunge", *New Media*, February, p. 35.
21. Loader, "Reconstructing an Australian School", p. 10.
22. Loader, "Reconstructing an Australian School", p. 11.
23. Stager, "Computers for Kids – Not Schools", p. 42.

As Gary Stager comments, this is what happens in an active Logo environment; you "speak French to the computer" and you get feedback. It's a take-off point. It gets to be addictive. You get hooked in a way that kids were once hooked on books.

No doubt there will be some "bad news" stories with electronic education. No doubt there will be some students who are resistant, and some who will find the task difficult. But such eventualities are challenges to be overcome; they are no more an indictment of the medium than the reality of a comparatively few non-readers was an indictment of the book.

Teachers, policy-makers and parents who persist with the idea that computers are alien, that they lower the standards, that they are bad for kids and an impediment to serious learning, are making a statement about their own inability to appreciate the significance or potential of the new medium.

Changing Teachers

One teacher at a conference publicly threw down the gauntlet (and had the sympathy of many of his colleagues), stating categorically that *he* wasn't going to allow computers in his classroom until it was well and truly established that computers did not corrupt kids. He was doing nothing other than the monks had done when they held that students who read the "pagan" classics were endangering their immortal souls. And with comparable effect!

Such a teacher is by no means an isolated case. In fact, I am constantly shocked at the level of ferocity and hostility that surfaces among some teachers when the introduction of computers is discussed. One head of science was adamant that computers had nothing to do with him: everything he wanted to do he could manage with graph paper, and it was a lot cheaper. It would be over his dead body . . . etc. Then there are those who shake with anger at the very prospect of computers, rather than books, as an information medium in the classroom.

So the problem of what to do with some teachers remains a thorny one nonetheless. Future generations of teachers (or resourcers, or back-up staff, or facilitators) will need no convincing of the new world that electronic information opens up. Perhaps we will just have to wait until they take up their positions. We cannot rely on all of today's teachers (in schools or colleges) making the necessary transition. There is much powerful and painful resistance within the profession. Sadly, like the monks before them, those who insist on the rightness of their principles and a print-based education will become increasingly irrelevant. Not "superior", as they would often like to present themselves.

It's understandable that many educators feel threatened by the new developments. They are right to think they are being deskilled. Iit is not a positive or pleasant experience to have your hard-won status, skills and qualifications dismissed, and your knowledge and values scorned. It will take an imaginative program to win such educators over to the new way of working in the world, so that they adopt a state of mind that prompts them to seek new skills. It calls for the establishment of a safe environment and massive in-service training. At the moment there's little evidence that this is on the national educational agenda, let alone taking place, with the exception of schools like MLC Melbourne. Darwin High School also has an exemplary computer program in place and every effort is being made to help the staff retrain and adjust to the new medium.[24]

As these schools already demonstrate, computer-mediated education will be a whole new world to teachers; one where teaching is subordinated to learning. Teachers will not have the status of being a knower, an authority figure, the active agent in the process, but will be called on as back-up for the questing students. The task of staff will be to serve student needs, rather than impose teacher dominance.

Apart from the traumas this role reversal will cause for some educators, there are underlying aspects like school organisation and classroom management which will also be transformed. For example, teachers will be teachers of human beings, rather than teachers of a particular subject (something which has been associated with the teaching of infants in the past). The new era will usher in a phenomenally different form of classroom "discipline".

For the past century or more, teachers have "controlled" classrooms by virtue of their monopoly on information. They have possessed something that students have needed – to get on in life, to pass exams, to get a good job. They have had the power to declare the student right or wrong. This has been at the core of the pass/fail system; it has been one of the means of maintaining classroom discipline.

It won't work in the future. Not just because kids will no longer put up with the cant, as they are lured by the carrot of the good job (the young are not so gullible any more). But because what the teacher knows won't count. Who needs all that stuff that was print-based when there is access to the limitless and seductive possibilities of cyberspace?

The relationship between teachers and students will enter a new phase: one where cooperation rather than hierarchy sets the terms. While no one would want to claim this as the panacea for all educational ills, it has to be an improvement on the covert and confrontational messages that so many institutions have conveyed in the past: that there are those who know and those who don't.

No aspect of education will remain untouched by electronic information. The emphasis will move (and has already moved) from the teachers and what they are teaching, to the students and what they are doing. One can only guess what this will mean in terms of student initiative and confidence. But it is going to be a very different picture from the one we have been accustomed to.

That students learn what teachers teach has been a cherished assumption of the traditional education system, despite the fact that overwhelming evidence to the contrary has been around for years. There has been a thriving branch of research

24. For further discussion about issues of access and equity, see pp. 249–51.

associated with "the hidden curriculum" which has left no doubt that, while teachers thought they were teaching anything from the history of Australia to evolutionary theory, the students were learning something quite other. And much of what they were learning was negative, and about self: they were learning about dependence, failure, stereotypes, and submission to authority. These qualities are not a desirable product of a professed egalitarian and liberating education system. (And they do not make for a well-skilled workforce, either.)

As yet, no one has done any longitudinal or systematic research on the impact of computers and constructivist learning, although it has to be said that the overwhelming impressions of those involved have been very positive. But one thing which can be predicted with confidence is that some of the old lessons of the hidden curriculum will no longer operate. Students will be learning very different lessons about the nature of knowledge – as well as a very different view of authority and autonomy – as they take charge of their own intellectual pursuits and call on teachers only when their assistance is required.

And there will be in this a "hidden curriculum" for teachers.

It won't be just the role of the teacher that changes. Everything from age-based classes to chronological progression and content-based curriculum, through to subject-oriented education, class periods, and examinations, is on its way to being restructured out of all recognition. And teachers (and teacher-training institutions) will have to adapt to these changed circumstances.

Teachers can't make these changes on an individual basis. Vast resources need to be committed to retrain and reskill them at this very moment; if technology can be defined as anything that has been invented since you were born, then most teachers are confronting technological challenges on a grand scale, and need professional development courses of comparable proportions.

As David Loader has commented, this "is not about training teachers to use computers in specific ways so they can train their students to do likewise". It is much more fundamental and far-reaching: it will change the "world view" of teachers; alter some of the belief systems and values that constitute their humanness; and place them in a different cultural milieu which is like migration to another country. It will change their relationship to students, subjects, and teaching. Teachers have to learn "to stand back from their practices and see which elements are still their exclusive responsibility and which the students can assume," says David Loader.[25]

Helping teachers to cease teaching what they were taught – that's the revolution. For some teachers who have experienced this change in the professional culture (at places like MLC, for example), this is the most exciting time to be in education: they are exuberantly rethinking what they know and what they do; they are taking up the challenge to be reskilled. But there are still many "dinosaurs" who do not want to

---

25. Loader, "Resourcing the Future", p. 33.

change and who are being given little or no opportunity or support for adaptation.

After I had delivered an address at a primary school conference in which I presented an electronic scenario and concentrated on the positive features for kids, the mood confronting me was sullen and suspicious. I am sure one teacher spoke for many when he said:

> It just looks like child-minding in a video arcade to me. I find the whole thing abhorrent. And I for one will have nothing to do with such a lowering of educational standards.

Cyber-education: Positives and Prejudices

For lots of reasons, not all schools will be able to make the dramatic shift that MLC has; but this should not prevent any school from starting to move in this direction. (Darwin High School has a policy which has been developed over the past two years and which is ahead of schedule. There is still a long way to go before all the students have personal computers and their own knowledge system, but it is a school which readily demonstrates what can be done.) Already, the opportunity is there for schools themselves to be genuinely multimedia: to include electronic information along with the print-based traditions, and to do it in the spirit of open-minded experiment. There is a growing range of multimedia products available to assist in the process.

In the United States, for example, an increasing number of educational authorities is exploring the multimedia database as the medium of learning. (It has to be said, however, that this is sometimes seen as bowing to the inevitable rather than embracing new and exciting possibilities.) The educational theory of constructivism which goes along with electronic learning is also gaining ground. It seems that teachers are becoming more at ease with "a system of inquiry in which the student is immersed in hands-on, real world situations and asked to provide the solutions".[26]

Besides, the students are happier. Many no longer feel so alienated. The discrepancy between their world and the educational world is not so great and the dissonance is diminishing. It stands to reason that kids who are stimulated and engaged are not so tempted to disrupt the classroom.

There is no doubt that schools will turn to multimedia – the issue is the degree of willingness. In contrast, the conventional book publishers are receding from the scene. In their place come the multinationals (which like Paramount and Time Warner have a media component) making some amazing programs. These include for example, *Pilgrim Quest*:

> Sold in both a home CD-ROM version and a videodisc aimed at the school market, *Pilgrim Quest* is a survival simulation in which children make decisions about settlement sites, planting, fishing and trading with the natives [*sic!*]. In addition to posing challenging math and social problems, *Pilgrim Quest* includes video footage especially shot for the application

---

26. Stansberry, "Taking the Plunge", p. 30.

that recreates Pilgrim life, as well as animation and video game exercises in such pursuits as hunting and fishing.[27]

As this quote indicates, the new media don't always change the old mindsets. Note that the children can negotiate with the "natives" – which makes it very clear that the white, colonialist, "English" point of view predominates. Moreover, most of the activities described originate in the roles and interests of men; nothing here about cooking, cleaning, or child-care. Or even about education and communication.

This could be a major area of future concern. Not only might there be just *one* version of the news in the global village (CNN); there might also be only one version of cyber-education; white, male, and Californian in priorities and perspective.

The potential for cultural imperialism, for propaganda and indoctrination, is enormous and insidious. This is a good reason that there should be a multiplicity of databases available. This is where Australia could make a great contribution.

Currently we are dazzled by the new packages that are coming on-line. Here are three examples with widely differing emphasis. In one module of *Science Essentials*, "students study the complex interplay of light and sound during a thunderstorm", and the storm sequence can be "stopped at key points while the students measure the amount of time between a thunderclap and flash of lightning [to] calculate whether the storm is approaching or receding, and at what rate and distance".[28] *Super Story Tree* is a creative writing tool on a floppy disk that allows students to create interactive stories using text, illustrations, sound effects and music. *Race to Save the Planet* uses video footage from the WGBH "Nova" series to investigate such issues as water and air pollution, the greenhouse effect and waste disposal. This interactive version teaches students the interrelationship between environmental and economic decisions as the student plays the role of the president of a juice factory.

And there is the *Aboriginal Encyclopedia*[29] – where the potential for CD-ROM to represent the multimedia nature of Indigenous society is quite extraordinary.

And these are but a few of the titles available!

No wonder the students think it is great:

The kids really take to the technology [says Mac Carey, who is working on *Science Essentials*]. Kids who feel shut out by a textbook or by a teacher can gain a lot from multimedia. If you use the whole program, including the full option science kits – the hands-on stuff – it takes all of their senses and brings them into the learning process.[30]

The students agree; "It's faster than flipping through a book", says one Californian fifth-grader.

---

27. Stansberry, "Taking the Plunge", p. 33.
28. Stansberry, "Taking the Plunge", p. 33.
29. *The Encyclopedia of Aboriginal Australia*, Aboriginal Studies Press, 1994, 2 vols and Apple Macintosh CD-ROM.
30. Stansberry, "Taking the Plunge", p. 33.

What we need to keep in mind, however, is that the involvement of the school (or the college) could be only a stage on the way to cyber-education. In the future, students will not have to attend a physical place in order to get information or credentials. The education scene will resemble Cyberion City (see pp. 96–8) or an extension of the distance education we are already familiar with. This will surely be the shape of the future – particularly if teachers and educators don't make the shift to the new medium. Mac Carey, a Californian researcher on multimedia projects, says he looks forward to the day when networked multimedia systems will allow students access to a library/database that contains text, video and sound, on-line. The student classroom could be the virtual world.

This could make educational institutions superfluous.

There are always losses associated with change. It is possible that some of the "high culture" from the past will go (just as some of the religious commentary from the manuscript period gave way to the pagan texts). But if we did have a readily available history of scholarship – of what counted as highly prized and essential information in one era, only to be discarded or even discredited in the next – we would have a much better basis for contextualising some of the changes we now encounter. And then perhaps the following altercation would not have taken place:

Teacher: What do you mean we could be studying Shakespeare through videos and CD-ROMs in future?

Dale Spender: Well, to some extent we already are. There are lots of schools these days where students' experience of Shakespeare comes mainly from videos – often of more than one performance. Where they can study style, make comparisons etc. I think this is fine . . .

Teacher: Well that's not my idea of studying Shakespeare!! It's the *text* that is important. It's going over it word by word that you learn what it's all about.

Dale Spender: I doubt if Shakespeare would have seen it in those terms. In his day, it was the performance that mattered. Maybe we could be getting closer to the real thing with videos and CD-ROMs, rather than . . .

Teacher: I've never heard such rubbish in my life. It's always been the word on the page that's been the test of greatness. And now you just stand there and say it's going to disappear . . . That we are just going to have bloody videos. I don't have to stick around to listen to this bullshit! . . .[31]

Clearly it's very hard sometimes to give up the idea of the primacy of print.

On the other hand, there can be no doubt that we will lose some of the good and the bad aspects of the traditional education system. We should be allowed to mourn the passing of these cherished icons. But at the same time, this should not stop us from seeing the advantages – and the inevitability – of the electronic age. We can always take comfort in the fact that human society did not end with the invention of the

31. I have chosen not to name the location where this interaction occurred; it could reflect on a particular school and reveal the identity of the teacher – which would serve no good purpose.

alphabet, or the arrival of the book, or the demise of Latin, despite all the predictions to this effect. And human society won't end with electronic education.

It will be a different sort of society; it will have different information and different values. As the generation caught in the middle, we will know about the old and the new. And the way change takes place from one medium to another.

Not everyone was in favour of the computer at MLC when the proposal to transform the system and go over to computers was first canvassed. And not everyone stayed. But the support of just about every member of staff, and the enthusiasm of the parents (and of course students) is now without question. There is even understandable pride in "the ladies with the laptops". And while MLC is a privileged place, it is not without relevance to other educational institutions. If one school can do it (and one *girls'* school, into the bargain), then others can too, as Darwin High School is demonstrating. The issue is resources:

> Only inertia and prejudice, not economics or lack of good educational ideas stand in the way of providing every child in the world with the kinds of experience [that cyber-education can make possible]. If every child were to be given access to a computer, computers would be cheap enough for every child to be given access to a computer.[32]

Electronic Scholarship

Status

It's not just the schools that are dealing with the challenge of the computer. The radical changes in teaching/learning, and in publication, are putting pressure on the traditional practices of universities as well. There is growing recognition that much of what has been valued and taught in the past has no place in the future (or the present). But there is no great conviction or consensus about what should be offered instead. Among those who are affected, there is much resistance to this aspect of the knowledge revolution.

Traditionally, universities have been "knowledge-making" places. A system of checks and balances or "quality control", in the form of scholarly conventions, has been designed to ensure that the knowledge is reliable. These conventions have covered everything from the methodology to the presentation of results; from scientific method to footnotes. But with the electronic medium, most of these ways of doing things are no longer appropriate.

Throughout the print period, the knowledge made in the university has been presented in standardised print format: research reports, theses, journal articles, etc. But this type of publication will be of decreasing importance in the electronic era. Already more and more knowledge-makers are moving away from print publication to

---

32. Seymour Papert & Cynthia Solomon, 1971, quoted in Stager, "Computers for Kids – Not Schools", p. 42.

television, radio – and, more recently, the Internet. It is only a matter of time before we have research reports and PhDs presented as software or videodisk. Even the printed journal articles will give way to electronic publication. And many of the present associated jobs will disappear.

There is no need for publishers, editors, reviewers – the gatekeepers and custodians of standards – when anyone can "publish" in an interview on the radio: or, with the necessary technology and a keystroke, can make public their findings to more than 30 million people in cyberspace. (In addition, they can keep updating them, which is much more sensible than the present arrangement; it could lead to the "never ending" PhD or the ongoing research findings.)

Many academics are beginning to wonder what they should do about this. They feel as if the rug has been pulled out from under them as they move from the stable and "checkable" knowledge of the print world to the "unreal" world of the electronic medium.

Academic knowledge, along with the academics themselves, has historically enjoyed greater authority and prestige than the information presented in the mass media. This is partly because of our cultivated faith in the printed word. If you read something in a book or a journal, it has to be true. In contrast, with the radio or the television – or conversation, for that matter – the spoken word has to be treated with a certain degree of caution and circumspection.

But now there's not just radio, television or newspapers. There's the Internet as well. And who knows what the status is of electronic bulletin boards? Who can determine the weight – or credit – that should be given to messages on e-mail? What veracity goes with announcements of breakthroughs on World Wide Web? This is the crux of the problem.

One reason we have readily accepted the reliability of academic information has been the process of *checking* that it goes through before it is made public. This has been an advantage of print and has made many of the technological achievements of the past centuries possible. Print stands in stark contrast to radio and television, where expression is spontaneous and instantaneous: speakers can say anything that comes into their heads. And some presenters do (just listen to a few of them!).

Traditional academic knowledge, however, has no such spontaneity; it has been subjected to scrutiny and assessment by experts before it is allowed to see the light of day. It is this "refereeing" process in which academics take such pride that has been responsible for much of the past public confidence in academic publications.

Of course, it is precisely this process of checking that will disappear with the shift to electronic media.

Credit

The drift away from print and towards the electronic has been going on for at least a decade. The pressures have come from all quarters, as academics have been called upon to be more accountable to the community which provides the financial support for their research. The Australia Council, for example, in 1990 published the booklet, *Unlocking the Academies*,[33] which urges academics to stop speaking only to themselves and to "go public" by making much more use of the electronic media. Many professional associations – scientific ones among them – are urging their members to explain the nature and significance of their research to society.

It has to be said, however, that most academics aren't too keen on informing the public by this means. The "ivory tower" syndrome operates; and there is also the belief (by some) that academic freedom means the freedom to pursue any research that takes your fancy without having to be "accountable" for it. Another reason is the deep-seated conviction that the public has neither the authority nor the ability to interrogate or to understand the nature of such grand intellectual activity.

A further reason for the opposition to publication by radio – or e-mail – is that academics don't normally get credit for their work when it is made public in this form. In a collegial community where status, success, reputation and promotion rely so heavily on the list of publications, any activity which doesn't count towards these isn't highly valued. An appearance on current affairs television may call upon as much expertise and take as much time to prepare as any paper for an academic journal, but this isn't usually acknowledged.

Preparing for the "live" publication isn't the end of the task, either. To confront cameras, to respond to an interviewer (not always sympathetic) and still to be perceived as credible demands at least as much skill in operating in the electronic media as writing an academic article required in the print medium.

But the strict criteria which have been developed for what counts as a publication have not been revised to take full account of the move from print to image. Some adjustments have been made, some "equivalents" established so that people in music, fine arts and performing arts, for example, can get credit for their work. But the process has to go much further; perhaps to the point where the print yardstick (increasingly absurd as the means of measurement) has to be abandoned. Authors are finding that the old, highly sophisticated system of copyright which gave them ownership of their efforts won't work for the electronic era; and maybe academics will have to reach similar conclusions about the old system of intellectual ratings, via print publication.

At the moment the academic community is generally trying to use one system (print) to measure the worth of another (electronic), and the results are often incongruous. For example, there is the problem of deciding how many exhibitions equal a book, or how many musical performances equate with a journal article. While the old print standard remains the benchmark for valuing non-print contributions, the academic enterprise leaves itself open to charges of being ludicrous. And most unintelligent.

The electronic medium will, of course, become a medium in its own right and will develop its own solutions. In the meantime, we are caught between two systems. But at

---

33. Australia Council, 1990, *Unlocking the Academies; Discussion Paper and Proposals for Action and Forum I following the National Ideas Summit*, 26 February 1990.

this stage, overall, academics who publish their work on the screen don't get the same credit as they would if they published in print form.

The view is that the new media don't count; but this is not the only reason that the academic community resists them. Probably the main objection to going public via the airwaves is that it lowers the tone and reduces the mystique of the academy. To go popular, to make an appearance in the mass media, to perform rather than publish, is predictably to be seen to "lower the standards". The argument against electronic publication (including publication on the Internet) is that it cannot employ the same safeguards and guarantee the supposedly high, reliable standards that have been the hallmark of the print publications of Academe.

Perform or Perish

From being granted credit for mass media performance, it is only a small step to being required to present *all* research in this format. This is where academic "publishing" is heading – with grave implications for academics.

When university teachers and professors find that it is mandatory that they present their findings in the electronic – rather than the print – medium, it is worth noting the pressures that they will be under. Their reputations as scholars will depend on their ability to perform, to make their work known orally. Their "image" will influence the rating they get for the purposes of appointment, promotion, tenure, etc. Just as it was essential to publish in print when it was the primary medium, it will become imperative to publish electronically when it is unchallenged as the dominant form.

There is some evidence that the shift from print is already under way. Harvard Medical School, for example, is encouraging its faculty to publish less – not more. Some universities are stipulating that candidates for appointment or tenure should select their five best papers as supporting evidence, and that the rest of their assessment will depend upon interview performance.[34]

For many university teachers, this emphasis on appearance could spell disaster. As a group they have not usually prided themselves on their standard of dress, or even their ability to communicate; to have to perform for their ranking will – to many – be little short of an abomination. But "no screen test, no job" could well be the terms of the academic contract of the next decade.

(Already at the University of Wollongong, faculty are being coached in the art of television presentation: "SBS TV executive Dr Austin Steel and . . . Mr Ian Hamilton, have been appointed honorary senior fellows to advise on setting up and running the joint Wollongong/SBS consortium which will provide fee-paying graduate courses via TV programs and videotapes.")[35]

Even if they are willing to perform or perish, the chances of success for many

---

34. Jennifer Kingson, 1989, "Where Information Is All, Pleas Arise for Less of It", *New York Times*, 9 July, p. E9.
35. See Eric Aubert, 1993, *Campus Review*, 15–21 April, p. 1.

academics could be slim. What will be the fate of the mumbling, shifty-eyed professors whose very presentation of their work, would – like Richard Nixon's five o'clock shadow – undermine the credibility of their research?

If academics cannot satisfactorily perform their own work, will they be permitted to hire someone else to perform it on their behalf, just as those who could not write in the Middle Ages had scribes to represent their ideas?

If the work is not performed, will it be deemed unpublished? This idea is not as far fetched as it might at first appear. We are rapidly approaching the stage where it is not enough to do the research, or write the book – the work has to be performed as well. The many writers "on tour" can testify to this.

Readings are an increasingly popular activity. Writers' festivals are now regular events on the literary calendar, as performance challenges print.[36] Instead of books being the original on which the performance (or film) is based, there are now instances where the book has followed the performance. *The Piano* is one example.

Performance is not confined to going public with the work. It is also reaching into the process of research itself. Action-based research for instance, illustrates the way in which market research, television and polling research are merging with the traditional academic model.

Those who have been reared with print are sophisticated researchers when it comes to interrogating a text, and critiquing premises and arguments on the printed page. But a new model of research is developing where the researchers interrogate not the text, but the author.

The book *Backlash* (Susan Faludi, 1991) provides some excellent examples of this process. The researcher/performer/author went to the producer of particular information and examined the context in which it was generated.

For example, Susan Faludi questioned the female author of many of Ronald Reagan's policies which firmly placed women back in the home. She found that the policy writer herself was a public and powerful woman, who had left her own husband in another state, looking after their kids. Susan Faludi had no need to critique the arguments. Like "The Investigators" of the popular television program, she "performed" the research.

On the one hand, the computer will mean that much more research can be undertaken at a terminal. Much more can be published electronically by means of a single keystroke. But as we move from page to screen, from print to image – as we move into the realm of virtual reality, for example – more and more emphasis will be placed on image and performance. Academics are not well prepared for this scenario.

From Peer Review to Newsworthiness

The traditional method for the publication of research has of course been print – which has had its own characteristic form of assessment. This process of checking and

---

36. See Dale Spender, 1991, "Introduction", in Spender, *Heroines*, Penguin, pp. 2–28, for further discussion of the shift from page to performance.

validating academic knowledge is known as peer review, and has been the widely established practice throughout the print period.

After many long years of testing and apprenticeship in the education system (high school, undergraduate, postgraduate student, etc.) and after a period of rigorous research, scholars would write up their findings, often under the supervision of experts, and then submit them for examination. This could take the form of sending off the article to the prestigious academic journal in the field. If the journal editor(s) accepted that the knowledge was appropriate for publication, the paper would be forwarded to reviewers, who were the professional international experts. Or so went the rationale.

The reviewers would vet the article; they would check the references, the footnotes, the sources and the acknowledgments – all the signs that the standard procedures had been followed. They would assess the article to determine how closely the researcher had observed the principles of scientific method – and got to the truth. If they were satisfied, they would vouch for the findings: the paper could be published, and the author given credit for the work.

This was how an academic reputation was forged; this was how the scholar was recognised and promoted. Theoretically speaking.

Researchers "owned" their findings, which when cited or used by others had to be acknowledged, referenced, footnoted. ("Citation systems" have been designed which count the number of times a scholar is cited in the works of others, and such figures have been included in the making of reputations. Yet there are obvious flaws in such a system: it has no concern as to whether the citations are positive or negative; and a group of people in a field could agree to cite each other with great frequency and mutual benefit.)

Sources could be traced; scholars could be questioned; professional societies could confer; the right people could be given prominence – and information could be trusted. This was the sort of accountability the academic community wanted – that of being accountable to one's peers; not to the public. This was the rationale behind the system. Because it was monitored by "experts", there was not supposed to be any room for deception or ruse; there was no place for the work of a charlatan with such a system of surveillance.

In real life, this method of journal review wasn't as objective as it was held to be; in retrospect, it was no more a guarantee of truth and objectivity than the Bible had been.[37] But it did involve a measure of "quality control". And there was a form of accountability. Academics, for whom reputation was so vitally important, had to verify, validate, and stand by the knowledge that was produced.

Of course, this didn't preclude findings from becoming discredited. Apart from cases of deliberate fraud that have been exposed, new research could call into doubt

---

37. See Dale Spender & Lynne Spender (eds), 1983, "Gatekeeping; The Denial, Dismissal and Distortion of Women", *Women's Studies International Forum* (special issue) 6, 5, for further discussion of the issue.

something that had previously been given the peer group's seal of approval. But this was regarded as an advance in knowledge, rather than an indication that the system might have got it wrong in the past.

This orderly practice endured for years. But with all the people and the processes that were involved, often many months went by between the time of the research and writing up, and the publication of the findings to the world.

Now this old and time-consuming procedure is being challenged and changed. Where once the discovery of DNA was reported in *Nature* after the work had been examined and vouched for, over a period of time, by peers, nowadays the latest scientific discoveries can be "announced" on the evening news. Once, the breakthrough was accredited to a particular person (not always the right one, as the case study of Rosalind Franklin makes clear)[38] and assigned a particular date in the journal; these days science goes public in very different forums. The latest promise of HIV research, the latest treatment for breast cancer – even the latest discovery of a gene for pacifism or feminism – get their first public airing as a dramatic presentation on the world news.

The content of news and current affairs programs reflects this shift as more and more items are information-based. Academic economists predict the highs and lows of the financial markets; Middle East specialists comment on the peace moves, criminologists use their research to show why jails won't work, social work theorists provide the data on the issues and incidence of sexual abuse, and engineers quote the studies that prove that the degree of radiation is not harmful and that the pollution effects are minimal.

Who knows how reliable this information is? Without the process of peer review and the safeguards it embodies, who knows whether the information is true? It doesn't have to be a case of a researcher trying deliberately to mislead. It can be that the researcher has missed something, left something out, deficiencies that the review process would detect.

In their survey of media-savvy 16- and 17-year-olds, the McCann Erickson researchers found that this generation was highly sceptical of information that was presented on the news.[39] They understood that there was no way of telling whether or not the news was intended to deceive. They even went so far as to compare the news unfavourably with commercials, which they rated as straightforward and reliable. But they were very aware that there's no "Advertising Standards Agency " for news broadcasts.

They were right to be suspicious. Without the aid of reviewers, references and reflective time to check out the report, there is no way of giving the academic seal of approval to news bulletin "publications". And some scientific claims that have been made and broadcast to the world have turned out to be very "false" indeed.

---

38. Rosalind Franklin was the woman who made a significant contribution to the discovery of DNA but who was left out of the recognition process.

39. See David McCaughan & Greg Wrobel, n.d., *Out of the Real: Teens and Technology, An Investigative Research Study*, McCann Erickson, Sydney; see pp. 107–8.

Sometimes unprincipled researchers, seeking a wider reputation (and remuneration), "exploit" the media with research reports that don't live up to their claims. Sometimes the media tries to use the findings of the research community to create a good story which leads to inaccuracies and misrepresentation. In her book *Backlash*,[40] Susan Faludi presents a superb coverage of the way that a male-controlled media selected, interpreted and used research findings about women to convey a preconceived image; no matter what the researchers found, the media could use it to discredit feminists and to foster the premise that women should be in the home. While some such presentations were distortions, others were completely in defiance of the evidence – but they fitted the editorial policy and made a good story.

Either way, the boundaries between academic expertise and the mass media are being increasingly blurred. The distance between reliable information and a beat-up is narrowing, as academics themselves are being required to go public with the aid of the new technologies which have none of the apparent protections of the fixed print medium.

The response of the traditional academic community to the mass media is not all that different from the response of the monks to the medium of print. In both cases the central concern is that information is being made public without their influence or control. For monks, it was the appearance of the book – and the publication of just about anyone who could read or write! For the conventional academics, it is the emergence of television and the publication of just about anyone who is telegenic which is being seen as a fundamental threat to standards.

This is even before we think about the implications of electronic publication on the Internet.

One issue that is raised which could be of interest to the ethicists and educationists, is in relation to the sort of society we are now becoming: whereas during the Middle Ages the most revered and reliable information was the *oldest*, these days it is the *newest* information which is considered the most accurate and desirable. This of course means that the pressure to update, to give credibility to only the latest information, makes us a community built on continuous change, rather than conservation.

None of our theories or practices have adequately equipped us for this new system of values. The closest model we have – according to Judy Marcure – is that of the troubadours and minstrels in the pre-literate period:

The messengers and minstrels who passed on information before the emergence of written communication often relied upon improvisation to make their information more interesting, in a sense "performing" the information, and accuracy was often sacrificed by these ingenious purveyors of information. Oral societies preserve knowledge in spirit, rather than according "to the letter" – a telling expression which reflects the pervasiveness of print conventions and the degree to which they have penetrated our communication concepts. The audience also had attitudes to information quite different from ours today. Even the

---

40. Susan Faludi, 1991, *Backlash*, Crown Publishers, New York.

increasingly literate audience of the Middle Ages was more interested in a good story than in acquiring accurate facts – and not interested at all in the author.[41]

Such attitudes to information can seem outrageous to those of us reared on print. It will be difficult for us to adjust to new concepts of information and the value or validity of accuracy. And universities will be in the front line when it comes to the challenge.

Some people have suggested that the entire academic edifice will come tumbling down as the institutionalised practices move from the sedate, scrutinised and controlled publication, in *Nature* for example, to the instant release of information in cyberspace. That in the scramble for the most recent information, the academic system will collapse. That there's no place for reviewing, and the accompanying scholarly conventions, on the information superhighway.

Electronic Journals and Salons

It's a bit of a halfway house at the moment, as we try to transfer some of the traditional practices of print to the computer screen. Electronic journals are springing up, for example; while there is no definitive record (an impossibility in cyberspace!) there are more announced on-line every day. At the moment, though, it is fair to say that there's a lot of fuss being made about very little; all the discussion is not much more than an academic debate on the pros and cons of publishing the old print papers in the electronic network. Not that this is without value.

Peter Lyman argues that the Internet has already changed scholarly communication:

> In many scientific and technical fields scholarly publishing is less important than access to preprints on the network. If you have to wait for something to be published in a physics journal then you are not on the leading edge. . . . Thus scientists and scholarly associations in the US are not thinking about document delivery as much as replacing print with networked forms of scholarly communication.[42]

Significantly, Peter Lyman goes on to add that when he asks academics if the printed journal is still important, "they tell me it is important, but not as a means of research or teaching, it's important for tenure and promotion".[43]

Electronic journals still don't count in academic terms. The main grievance against them is more than predictable: *anyone* can publish in this format. All it takes is a stroke of a key. Far from being a scarce resource, information on the Internet is ubiquitous. Like the professional readers of old who found themselves deskilled once everyone could read, there are academics who are disturbed by the prospect of losing their professional status once the gatekeepers are removed.

Interim measures have been suggested as a way of keeping some quality control

41. Judy Marcure, 1990, "New Directions in Scholarly Communications", *Dynamic Documents and Print Conventions*, UNSW, p. 1.
42. Lyman, Follett Lecture.
43. Lyman, Follett Lecture.

– such as a form of sponsorship or "reputation verification", whereby distinguished scholars could attach their names to the article making it clear that they vouched for the standard of the information; that it had their approval; that it wasn't just *anybody's* contribution.

But even if some sort of peer review and imprimatur could be developed, the problem of electronic journals would not be solved. From the academic point of view, the inability to control copyright is even more vexatious. It's not just that *anyone* can publish on the nets, but *anyone* can take the information from the screen and make it their own with not much more than another stroke of the key. The academic community – which is better known for its tendency to hoard information than to hand it out free – doesn't find this a satisfactory arrangement.

"The principles of copyright essentially protected original works fixed in a tangible medium of expression", explains Jane Levine, a leading intellectual property lawyer, as she talks about "data soup": "Present copyright does not protect ideas, theories, facts, raw data, systems, processes, methods of operations, principles or discoveries."[44] Once their work is on the net, academics would have trouble claiming credit for their ideas (which can have considerable economic potential), let alone claiming credit for their publication record.

But all this discussion about electronic journals is purely academic. It points to the print origins of the paper-writers who want to transfer their old formats to the new technologies, and who find that they won't fit. Publishing "papers" will not be a priority in the future.

Already some interesting variations on the theme are emerging. The use of the electronic salon is a good example. The first electronic salon that I know of took place in the USA in 1993 when a community of women interested in the issue of women's communication on the net held a salon for a period of two weeks.[45] In that time the participants agreed to commit a certain number of hours to "conversation", and the result was exhilarating and, by all accounts, extremely satisfying.

Those who wanted to prepare and publish papers put them on the net. Soon there were wonderful responses. Connections started to be made, comments were forthcoming. All the information presented was in a constant state of flux – being negotiated, modified, "appropriated" and extended. Socrates would have given his blessing to this dynamic process and exchange of ideas.

And the women who participated found it very rewarding. Some women reported that they found the electronic salon much more useful and enjoyable than an average conference where they might not have felt able to question the presenters, and where there was little time to reflect and respond. They valued the ongoing nature of the process; they also indicated that the electronic salon gave them the sense that they were not individual knowledge-makers – but part of a team.

---

44. See Trudi McIntosh, 1994, "Copyright Chaos in 'Data Soup'", *The Australian*, 12 April, p. 24.
45. For a more comprehensive coverage, see pp. 239–41.

The Art of Information

While most members of our society don't go round talking about the nature of knowledge, the changes that are taking place in academic information are having profound effects on all of us. Like the people of the Middle Ages who had to adapt to print, we are moving into a world where our old certainties are no longer available. We can't believe in the old faith anymore.

Throughout the manuscript period, the Church had taught that there was only one truth – God's. But with print, and the rise of the secular university and the scientific revolution, this truth was challenged and changed. There are still those who believe the story of creation as it is recorded in the sacred texts, but most of our modern life is now based on the very different truth of scientific method. The science of the last centuries had given us a different explanation of the origin of the world, and the species, with the theory of evolution.

Now, however, the scientific truth of the print period is itself being undermined as we move into the electronic era. The long-prevailing premises of objectivity, empiricism and truth are all on the way out. We no longer talk about the one and only truth; these days we have many truths. We think in terms of pluralism, in terms of transitory truths, rather than of the fixed, focused, finality of print.

Earlier generations might have believed in the concept of true and false; they might even have been confident that they could tell the difference between the two. But future generations will have no such luxury. We are beginning to learn that there are many truths, and that the idea that there is only one is no longer meaningful. (Maybe this is a case of "back to the pre-print" values?)

We are starting to think about *good* information, rather than *true* information. And good information, will probably be like good art in the not so distant future. It will resemble a good play, a good painting, a good performance. Good information will be all of the above, and in multimedia form – something which we enjoy, and which enhances our life. Appreciation is what we will need to cultivate, rather than the simple criteria of right and wrong.

Just as we haven't expected art to *prove* anything in the literal sense of the word, so will we stop trying to get information that has been proven. At least, that's how the art of information should work.

Personally I must admit to feeling more than a little unease with the idea that there are no guarantees. What about air travel, surgery, electricity? I am not sure I can be so philosophical as to accept that the safety of the plane, the certainty of the surgery, the soundness of the electricity supply, are all a matter of opinion or appreciation.

But I do acknowledge the underlying principle that creating information will be like creating other art forms: what is considered good will be what is seen as skilful – and whatever takes your fancy.

To some extent, sound and certain information has always been something that you like – and that you can admire. This is why white, professional men have stood by the knowledge which they have made over the centuries and in which they have enjoyed such privileged status. It's also why there are those among them who are now looking for someone or something to blame, as their particular truth and standards begin to disintegrate.

Over the past two decades, the people whom scientific method for so long excluded have started to put forward their own explanations, theories and truths. Black Studies and Women's Studies are but two of the new bodies of knowledge which have appeared on campus and have insisted on their own authenticity. They have demanded that their version stand alongside that of the white male record, and that it be taken into account. They insist on seeing men from women's point of view; and seeing whites from the perspective of blacks. These unpalatable "truths" have not usually been there before.

Women's Studies (or feminist scholarship: there is more than one way to describe the activity) has challenged men's right to name the world starting with themselves. "Objectivity" is just the name that men have given to their subjectivity, said Adrienne Rich.[46] Thus she exposes the whole elaborate process of knowledge-making in which women were the aberration, the deviant, the footnote in men's record of their own achievements.

Hundreds, if not thousands, of feminist scholars have made a contribution to the exposé of the inaccuracies and inadequacies of academic knowledge over the past decades. From Jessie Bernard to Valerie Walkerdine, from Mary Daly to Donna Haraway,[47] they have identified the sexism both in the information itself, and in the way it has been put together. But they have not always been thanked for their efforts in making the knowledge more representative of humanity.

On the contrary, they have often been accused of being the cause of the erosion in academic standards. This is called shooting the messenger – when the message doesn't meet the approval of the powerful.

Because Black Studies and Women's Studies were the beginning of the end of the white, male truth in the groves of Academe, they have been on the receiving end of much harassment and vilification. Conventional scholars like Alvin Kernan and Allan Bloom, for example, have denounced the feminists who are supposedly responsible for undermining the great academic traditions of truth and freedom – as these men have experienced it.

To some extent, the white male response is justified. It *is* the end of an era, the end of an institution as they have known it. But feminists and blacks are no more a cause than the myths and legends of ancient Greece and Rome were the cause of the Church's

---

46. Adrienne Rich, 1979, *On Lies, Secrets and Silence*, Norton, New York, p. 208.
47. See Jessie Bernard, 1973, "My Four Revolutions: An Autobiographical History of the ASA", in Joan Huber (ed.), *Changing Women in a Changing Society*, University of Chicago Press, pp. 11–29; Valerie Walkerdine, 1988, *The Mastery of Reason*, Routledge, London; Mary Daly, 1973, *Beyond God the Father*, Beacon Press, Boston; Donna Haraway, 1987, "Animal Sociology and a Natural Economy of the Body Politic", in Sanda Harding & Jean O'Barr (eds), *Sex and Scientific Inquiry*, University of Chicago Press, pp. 217–31.

downfall. In both cases, the medium was changing and a climate was created in which new ideas – feminism, or the humanities in the earlier period – could gain a foothold.

What we have witnessed in the past few decades is the transition from the fixed to the fluid, from the monolithic model to the multiple one. As one would expect, there are complaints from the white male professionals who feel that they have been deprived of their status and skill as they are joined by a crowd of knowledge-makers who have equal claims to be taken seriously.

Issues for Scholarship

While at every level the process of knowledge-making is changing, not much attention has been given to the revision of scholarly conventions. Many of the practices which emerged last century still persist, despite their likely irrelevance to – or, indeed, their possible interference with – the presentation of information.

In many universities it is as if the power of the computer does not exist. Although it has dramatically altered the way information is processed, its role and influence have not been taken into account. The issues that the computer raises go beyond data-gathering and the administration of all these new requirements; there are ethical and educational issues to be considered also.

For example, the relationship between technology and scholarship demands urgent attention, as does the relationship between technology and scholars. In the present context in which so little investigation has been undertaken, some unquestioned and highly dubious assumptions could readily be made. Is the person with the biggest computer judged to be the best and most proficient scholar on the strength and capacity of the hard drive?

So much has changed and yet there has been little discussion about the significance of these new ways. It was once the case, for example, that in any research project (PhD included), the "review of the literature" was often the most time-consuming task. Finding out what other scholars were doing in different parts of the world called for considerable ingenuity. But these days, an electronic search and a data-surf of the bulletin boards takes but a matter of minutes, and provides more detailed and up-to-date information than two years of solid work might have yielded a decade ago. (Of course, how you use this information is a different matter; but that was the case with print as well.)

Yet this "revolution" has not been reflected in the established pattern of scholarship. PhD students, for example, can still find themselves bound to a minimum three-year degree, although a substantive part of the requirement has been reduced from months or years to weeks.

The reality is that the review of the literature might now have become a perfunctory exercise for some, but this hasn't led to any great discussion on the need for new and different demands in other areas. There doesn't seem to be a lot of scholarship on scholarship – which is quite astonishing in the circumstances.

For students are already seeking to present their work by electronic means. There is a perceptible and growing tension on many campuses as the younger generation develops software, and undertakes pioneering and innovative work on videodisk or the Internet. The traditional academics cling to the view that it is print and publication that counts, and if the students want to pass, or to get appointments on the faculty, they will have to buckle down and get something on paper, preferably in a professional journal. This is the only standard that many established academics understand.

It is of course the traditional academics who are going to have to change their ways. It is to be hoped that they will go graciously, and that there won't be any student riots on campus. But many of those who currently sit on dissertation committees, appointments boards and tenure hearings will have to abandon the standards that they have lived by and which are at the core of their professional and sometimes personal identity. It is understandable that they could feel completely deskilled by such an eventuality.

But ignoring the changes and the pressures won't make the new medium go away. There has to be a reckoning, sooner or later; we have to start thinking about the implications of scholarship when students want to do their dissertation as software or video; when they want to present their research as virtual reality; when they want to collaborate, to work in teams, and to collectively process information, rather than to claim individual ownership of it. And when the following issues will be raised.

Competency with the Medium

By the time students get to postgraduate stage, it is assumed that they are competent at reading and writing. It is a condition of their success. They have come through a system based upon literacy skills. (If they weren't good at writing and reading, they could have done a communication course in the English Department ostensibly to improve their competence.)

- What will be the equivalent skills as we switch to the electronic media? When students want to work on-line, prepare their assignments as software, do their theses on videodisk?
- If students are required to work in these electronic media – or if they seek to do so – should we start demanding some proof of competency? If so, who will administer this system? Who will teach it, set the standards, develop the course criteria?
- Should the Media Departments (i.e. the former Audio-Visual Centres) of universities fulfil this role in much the same way as English Departments did in the print system?
- Where would this put Media Departments? They would soon be the most powerful departments on campus. Would they be in the merged computer/library centres? Would they be part of the Education Department?)

There are numerous issues here in relation to university organisation, curricula, and the support of students and staff.

Relationship with the Media Department

- Does the student produce the software/video alone? Or within the Media Department?

- In print terms and times we were content to allow students to hand over their material to a professional typist for presentation; should we allow video students to hand over their data to the professionals for presentation?

- What is the role of the students in the teamwork of video production? Are they interviewers, presenters, editors, scriptwriters, directors – or technicians? Do they hold the camera? Or is there a dividing line (which would be extremely difficult to determine) whereby the students provide the intellectual content and the Media Department does the hands-on work?

- What level of professionalism is required? A handwritten PhD thesis would not have been acceptable in past decades (although it would have been in the nineteenth century); is "home video" unacceptable as the standard for research purposes?

Supervision

- How relevant is the concept of academic supervisor in the creation of software and videodisks? And how many old academics will be able to learn new tricks?

- Will we have a generation of students who have no electronic mentors or supervisors, as print gives way to the electronic?

- Or is this the beginning of team supervision: an electronic "committee" containing technical experts as well as scholarly ones; a partnership in which the supervisor is part of the production team? (There is a model here in the old print publications, with junior authors.)

- What is being supervised? Content or form? Medium or message? Quality of ideas or of representations? (To what extent did the quality of the prose determine the evaluation of a print PhD or research report?)

Audience

- Who is the video or database for? What are the criteria in the infotainment industry? Is it sufficient to present the information – or must it grab the imagination of the audience?

- Who is the audience: the examiners, or an on-line audience? It has been argued that one advantage of videos, etc. is that they can reach a wider audience.

- If the examiners are not the audience, then what are the criteria for excellence? Perhaps two versions are necessary? This is comparable to the dilemma that often confronted the print PhD candidate; there was one version for the examiners and then a rewrite for publishing purposes.

Ownership and Intellectual Property

- Who will own the video? The issues of intellectual property are overwhelmingly complex in any area of electronic information – but in relation to the student and the university, they are mind-bogglingly difficult.

- Will the Media Department which supplied the equipment, the personnel and the facilities hold the copyright? If so, can the university sell or distribute the database (or parts thereof) or market the video?
- Will this be the business of the university of the future, selling the information produced by students?
- Can the student who has had full use of university facilities to develop software, videos, teachstations, etc. retain the intellectual property rights and sell them to the highest bidder? (Consider the example of the University of Illinois, where the designers of Mosaic have set up their own business.)
- What about the issue of teamwork? How many students can obtain a degree from their collaborative efforts on one piece of work? Who will own what on the team? How will it work for promotion, tenure, etc.?

Raw Footage and Data

- What about the role of raw footage, the roughs or drafts of the video, in this context?
- What is the role of the agent in making the selections or doing the editing?
- Will students and researchers be required to retain the raw data, in the same way as engineering students are asked to provide background data not included in their dissertation, for example?
- Will it be necessary to check what the student has left out?
- What about balance, bias, comprehensive coverage? Are we going to have occasions when students are asked to supply raw footage and are then disqualified on the grounds that they have distorted their results?
- If all the raw footage is to be preserved, who stores it? How is it filed? Who else can have access to it? Who is in charge?

Script

- No video or multimedia package could be put together without a script, a working outline of the contents and structure. Should the preparation of this be part of the research process?
- If it is to be included, how should it be assessed?
- Or is the whole idea of a script just a print hang-up? Perhaps the "doing" will soon be primary, will exist in its own right, and we won't need the outlines and forms of preparation that are integral to the fixed, print mode.

Data Gathering

- How are data gathered, and what are the issues here – particularly in relation to privacy?
- If video footage is being made of child sexual abuse or some aspect of Aboriginal life, for example, what rights do the *objects* of research have? Are their faces blocked out? What sort of permission should be signed?
- Who would administer such a system?

Distribution

- Will PhD videos and software gather dust on the university library "shelves" as happened with many print PhDs, or will the new forms be in demand in a way that the old ones were not?
- Could researchers and students stipulate that videodisks be on restricted use, for research purposes only? Or will the Library or the Media Department be able to distribute, or sell file tape, etc. as a legitimate part of the information business?
- How stable will the product be? Can anyone add, subtract, cut and paste the original?
- Who can copy? And how will fees be collected? (In Australia, the Copyright Agency Limited, which collects the fees for photocopying in universities, is working on a system to collect licence fees for all electronic data.)
- Who decides what can be sold?
- How many intellectual property lawyers will the university need to employ?

Methodology

- Is there any accepted methodology in the new medium? While the computer can't do everything, it will still be the major source of research activity and information. What effects will this have on methodology? Will there be more innovation – or less?
- What are the replacements for the conventional signs of scholarship and reliability? What is the electronic equivalent of the footnote? How are sources to be listed, other works cited?
- What is the meaning of originality (when so much can be "copied"); and what about excellence?[48]
- Will there be a controversy about the distinction between educational material and amusement value, between information and entertainment? Will research itself be dismissed on the grounds that it is boring and anti-educational?

When it comes to all the issues associated with scholarship, the conventions about standards and credibility which are at the heart of the academic enterprise, it is clear that there is a revolution going on. As with so many aspects of university life, the practice of scholarship and the "production" of research could well become unrecognisable as we move from page to screen.

---

48. I would like to thank Prof. E. T. Brown for his input into this section.

Universities

Virtual On-line University

In September 1994, the Virtual On-line University (VOU) became real, and challenged the model of universities we have had for the past centuries. The stated goal of this new global, electronic university is to use the Internet to provide low-cost, high-quality education, which can be accessed at home, at school, or at the workplace, anywhere in the world.

Since it was announced on the Internet, millions of people could have learnt of the existence of this new university. (The information turned up on my e-mail more than half a dozen times, redirected to me from different countries by colleagues who thought I would be interested.) VOU "opened its doors", metaphorically speaking, with a liberal arts degree and the expressed aim of soon providing postgraduate possibilities. Information was also included on its "facilities":

> Virtual Online University operates within a Virtual Education Environment using Multiuser-Object-Oriented environment database software (a MOO).[49] Some MOOs are programmed as virtual cities and research centers, others as educational environments. VOU will use various MOOs as online virtual campuses, including a traditionally designed university campus and one designed as a Geosynchronous Earth Orbit. Others are currently in planning, including an undersea environment.[50]

These are new terms, for new educational forms; they offer a variety of stimulating educational experiences; and they are open to anyone who can make use of them.

For those of us who are used to thinking of universities as places with highly restrictive entry requirements, the VOU comes as something of a shock. Anyone can enrol. All you need is the technology, not the paper qualifications which prove that you have passed certain tests and are eligible for inclusion in the quota system. There are no more "scores" or competition for limited places. The emphasis has switched from what you know to what you can do. (All you need is the wherewithal to do it. This is, of course, an issue of major importance, and needs to be urgently addressed. How are we going to make electronic access available to everyone in the community? When will computers and Internet connections become standard household equipment like televisions and telephones?)

Open Learning in Australia has – to some extent – helped to pave the way for the new era of on-line *open* education. Anyone who wants to can enrol in the courses provided by the Australian Broadcasting Corporation (ABC) in conjunction with a number of universities. Radio, television, video and audio tapes – and some Internet and e-mail facilities – along with traditional course materials, are readily available to

---

49. See pp. 96–8 for discussion of MOOs.
50. Internet, September 1994.

those who choose to participate. (And people are doing so out of interest – not just because of the prospect of qualifications.)

Edith Cowan University has also taken one of the first steps in the direction of the electronic campus. "Students and university staff are now linked through a computer network using Telecom Discovery software. They can send messages to individual 'mailboxes' or to 'noticeboards', and 'chat' during open forum sessions."[51]

There has been no great backlash about a "drop in standards" resulting from the Open Learning University. Probably this is because there aren't many graduates as yet. No doubt the criticisms will come, and there will be some who insist that degrees obtained by these open means aren't worth as much as those based on supervised entry and traditional teaching methods.

Of course, if the degrees that are granted to Open Learning graduates actually come from the prestigious universities involved in the courses, it will be difficult to pick the "open" entrants from the ones who passed through all the hoops.

Besides, qualifications are going to be much more fluid in future. Where you came from (be it via Open Learning, mature age entry, special entry category or as a school leaver) will be of much less interest than where you are going, and what you can do. And this will have ramifications for every aspect of university organisation – and philosophy. Already the greater emphasis being given to the identification of "prior learning" is indicative of the shift that is taking place.

Even the concept of a university degree – based on sequential combinations, graded units, specific content requirements, etc. – will become rapidly and increasingly inappropriate. Students will want to do courses on an infinite number of professional and recreational topics, no matter their age, or whether they are at home or at work.

The notion of the structured degree belongs to the pre-electronic era. It was based on the premise that you could be trained once, and that was it. These days, there is widespread recognition that learning is an ongoing process. And this is the challenge for universities.

Lip service might have been paid to the notion of "life-long learning" in the past, but the pressure of the electronic medium will make it a pragmatic reality. Learning throughout life will become the model; and the electronic university could become the means of participating in this process. With the entire community transformed into a "learning society" we will all need to be plugged into the Internet, to "play" with information as a work, or a leisure activity. (Anyway, who will be able to tell the difference?)

This on-line educational institution, with its open entry and boundary-less offerings, has enormous implications for universities.

For example, will we still need physical space as the university? Or will it be the virtual campus or the electronic campus that becomes the order of the day? And if we do need a physical space where staff are housed and students meet, just how many real

---

51. See *Campus Review*, 30 April–6 May 1992, p. 5.

campuses will be needed? Perhaps one will be enough, particularly if they all continue to provide much the same sort of thing.

Universities might have to follow the lead of advertising, and appreciate that there's no longer a mass market but a multitude of niche markets.[52] Instead of providing mass education, the role of the university will be to coordinate customised and individualised courses. Universities will no longer try to do everything. They will develop specific areas of expertise, much as they did in the Middle Ages – where the academies of one country would specialise in "multiplication" while those of another would become masters of "division".

Another possibility is that some of the contemporary universities won't be able to make the dramatic leap to function more like movie studios than lecture theatres. Despite the predictions that university administrators of the future could come from the ranks of media barons and movie moguls, a number of universities won't be willing to head in this direction.[53] And some that are willing won't know how to usefully go about it.

As the VOU, and other alternatives, start to attract students, Australian universities (and those of other developed countries, as well as some of the institutions of the developing world) will have to make radical changes if they want to be part of the infotainment industry, and provide services to the world's "learners" or "players".

Internet

There is no doubt that the Internet is going to be *the* educational hub of the twenty-first century. Once more we can see that the medium is going to change the message. Research and learning will not be the same activities that they have been with print.

Research for example will be – if it is not already – something that can be done primarily with a computer. Of course, a lot of data will have to be fed in and a lot of software will have to be developed – but then literally millions of people will be able to access, interact with, synthesise, simulate and "innovate" in the electronic medium.

(This will be true not only for the humanities: laboratory and practical work can be conducted this way as well. Engineers can use computer simulators to build model bridges and to see if they fall down; surgeons can practise anatomy; pilots can put in hours of flying experience; economists can run their figures; social workers can interrogate census data; and musicians can watch, listen to, and rehearse with orchestras from all parts of the globe – all without ever leaving their terminals.)

Through the electronic network, scholars will not only have access to the world's information for perusal; they will be able to communicate with each other. This has led Parker Rossman, author of *The Emerging, Worldwide, Electronic University* (1992), to describe this electronic community of academics as collaborators – individuals who are working in a "collaboratory". These global collaboratories "can

---

52. See McCaughan & Wrobel, *Out of the Real.*
53. See Guy Healy, 1994, "Media Barons Tipped to be VCs of the Future", *The Australian*, 19 January, p. 16.

increasingly mobilise many minds and thus facilitate a collective intelligence", he states – somewhat cavalierly.[54]

The idea that the world's scholars will be able to meet and work in a cyber-collaboratory, where they will freely exchange information, is a most attractive one. The prospect of putting an end to national boundaries and providing free information would win widespread support. But there are other factors which also need to be taken into account.

The first one is that in the electronic community, the *only* scholars will be *electronic* scholars. Anyone who is not wired up, who tries to undertake research by traditional means – in either a developed or a Third World country – will not exist for research purposes. This will result in a skewing of scholarship towards the technologically competent and equipped.

The bias extends further. Being on-line won't guarantee full participation in this "collective intelligence". While it is theoretically possible for everyone in the cyber-community to contribute on equal terms, this outcome is not likely.

Instead of drawing upon the diverse resources and perspectives of the global village, the odds are that the views of the most powerful will be imposed on the rest of the world. Not to put too fine a point on it, the concern is that the world will become Americanised, rather than that the United States will become more attuned to the customs and cultural understandings of other countries.

There is growing apprehension that what we could get is a new form of colonialism: one in which the perspective of the white Californian male predominates; an academic version of CNN.

Universities and scholars will play a different – and much more competitive – role in the global context. This fact will also put pressure on researchers to have their particular version of experience accepted and become "the rule". They will then have "product-lead"; they will be in a powerful and profitable position. Increasingly, universities will become like corporations as the information industry consolidates.

Universities will not be competing only with each other in the marketplace. At the same time as they are becoming like corporations, corporations will be on their way to becoming like universities. Both will be vying for a place in the information business.

There's no reason that companies like Optus could not become producers as well as deliverers of information. Communication industries, computer industries, advertising agencies, PR companies and polling agencies, along with document delivery services and, of course, computer games manufacturers,[55] will all be part of the infotainment sector.

Academics who are currently complaining about the erosion of their collegial

---

54. Parker Rossman, 1992, *The Emerging Worldwide Electronic University*, Greenwood Press, Westport, Conn., p. 49.

55. See McKenzie Wark, 1994, "The Video Game as an Emergent Media Form", *Media Information Australia*, 71, February, pp. 21–30.

conditions haven't really started to feel the pressure as yet. The old ideology of academic freedom, of the right of the researcher to pursue any subject of interest regardless of commercial value, has long since passed. In the near future, scholars who have no market for their research, no "buyers" for their information, will have no place in the university. They will be a luxury the institution cannot afford to carry.

The push for relevant and saleable research will also flatten the pyramid of scholarship. It will not be the highly elitist model that has prevailed in the past. Anyone who has access to the Internet will be able to contribute. They will be encouraged to do so in the forward-thinking institutions.

The trend towards the democratisation of research and scholarship (which parallels the democratisation of authorship) is already discernible. Research is no longer confined to the university. Market researchers, television researchers, etc. are making a significant contribution to the information industry (and often displacing the university researchers in the process).

Even within the university, the role of research and scholarship is no longer strictly limited to an academic elite. In the early 1990s for example, in one of the new Australian universities, it was a Human Resources Director who had the highest research profile (and brought in the most money). She specialised in the area of women and management, and the researchers in this instance were drawn from the administrative staff. This example shows how far things have moved even at this stage.

There's no point in traditional academics protesting that this new breed of electronic researchers are not and never can be scholars; that they have not served their "apprenticeship"; that they have no understanding of "excellence" and no allegiance to a professional association. There is no point in declaring that the new telegenic hosts of Open Learning are an insult to the very nature of scholarship. It's a bit like the scribes of the Middle Ages protesting about the introduction of the book and the new audience of readers that it generated.

The process of "knowledge-making" is increasingly located on the Internet. While universities will possibly be able to shape and influence some of the developments, they will not be able to stop or alter the direction.

Universities historically have had the dual role of undertaking research, and engaging in teaching. If there's no room for the traditional scholar in cyberspace, there's not much accommodation for the traditional teacher, either. One reason is that only a few teachers – and only the telegenic ones – will be broadcasting in the Open Learning context or on the Internet, which will drastically reduce the numbers required.

But there's another factor as well. As the switch is made from content to process, the expertise of many teaching academics will be called into question. With no discipline to teach, with no professional identity or allegiance (as we now know it) associated with history, or Japanese, or even physics, for example, academics will experience a crisis as they seek a new role.

Fear of losing one's job is only part of it. There will be plenty of jobs in virtual education: providing learning experiences on the Internet will be a labour-intensive activity. It could well take all the teachers we currently have in universities to meet the student need for facilitators to guide them through the databases and help them surf the net. But it is doubtful whether philosophers or chemists, etc., who have enjoyed the status and authority of experts in their disciplines, will take kindly to being support staff on the educational Internet.

Credentialling is yet another area of activity where universities could find a role. There is going to be demand for an agency which sells qualifications in competence, which is prepared to vouch for the skills of the engineer or the social worker. Of course, there won't be anything to stop corporations (or employers) from developing their own tests for performance, but they could license the university to provide this service for them. Not even this role is assured, however.

Universities along with other corporations are going to be competing for a fast-lane on the information superhighway. Their success, or otherwise, will determine which ones survive; and in what form they will continue. Without doubt, we don't need as many of them as we have at present providing the traditional print-based curriculum. As universities are restructured, it is clear that any academics who can't keep up will find themselves left behind. The pattern is here already.

As universities scramble for their niche in the global infotainment industry, they will cease to be the "public" institutions which have traditionally been charged with meeting local and national needs. The pressure of market forces could cause policies based on tenure, or equity, for example, to be abandoned. This could have consequences not just for the staff, but for the student body.

In the electronic age, information is the source of wealth and power. The stakes are high – and this is the game in which the universities are playing. Their past purposes and practices have not prepared them for the pressures they currently face. It can be predicted with confidence that the Internet University will be vastly different from the print-based model of the twentieth century. There's not a lot of time to get the house in order.

Virtual Education

Some schools are now experimenting with the "classroom without walls", as students make use of their laptops, and interactive multimedia packages replace textbooks. At this stage, there are a few schools where the students have e-mail addresses, where they are on the Internet or networked with other institutions. But these are by no means the norm. In general, school learning – even when high-tech – is primarily school-based.

In contrast, universities are now a long way down the road to the information superhighway or cyber-strata, and the virtual (and simulated) classroom is verging on reality. It will probably be but a short time before first-year students – no matter what their entry qualifications – will be issued with a "package" which contains everything from television timetables, to audio tapes, e-mail addresses, and passwords which permit entry to international databases.

For the university student, then, it will not be all regular classroom with four walls and an out-the-front lecturer. Rather, there will be the opportunity to become part of the learning society, at any hour of the day or night, by plugging in at the library, at home, or even at the workplace.

University teachers will still be around. They will be there to help students plan their own courses that have built-in feedback. But basically students will be in charge of what they want to do and the pace at which they do it. Independent self-directed study will be the new priority.

Students will meet in groups, of course, in this networked community. We have examples of how this is currently being done – and how it can be done with more readily available technology. At one level are the writing classes already established, such as those at the University of Illinois (Urbana Champaign). Gail Hawisher and Mary Hocks have developed state-of-the-art computer and composition programs which allow all the members of the class to see what everyone else is writing, on their own screen. This networked "hypertext" process is literally collaborative writing. Students are working on their own – while they work together. At the other end of the spectrum, geographically separated students from all parts of the globe can meet together in real time, with or without a teacher.

On-line students (like academics) are already able to set up class conferences, where students can meet electronically and exchange information. Many of them think that the new medium is just great. They can speak when, and as often, as they want while they are at the terminal. They might not get immediate answers, but their messages will certainly register on their tutors' screen, and they won't be neglected or overlooked in the same way that has often happened in the conventional classroom.

Clearly, the virtual classroom has many distinct advantages. When everyone can speak out without fear of interruption, shyness ceases to be an issue, according to those who believed that they did not get a fair go in the traditional classroom.

It could be that on-line communication provides the best opportunity yet for solving the problem of boys getting more than their fair share of teacher attention. When learning is student-initiated, rather than driven by teacher response, there's nothing to stop the girls from posting as many messages as the boys, both to other class members and to the teacher. If the teacher response is not dictated by the needs of discipline, the need to control and interest the boys, and to keep them from disrupting other students (and classes), there's no imperative for the teacher to concentrate on the messages from the males.

Except out of habit, of course. Or as a result of sexist bias which accords male experience more significance.

Each student will be able to control their own learning environment when they are on-line. They will all be able to make contact with their tutor (either directly or by e-mail). There's nothing to prevent teachers and students from building up a good rapport. You don't have to meet students face to face to deal with them "in person".

With on-line education, those who are in any way disabled won't need a wheelchair or assistance to attend classes. (Even the difficulties of blindness could be overcome; the voice-activated computer could be the solution.) You won't have to get to the university during office hours to see your teachers or supervisors; they are available on e-mail – even when they are on sabbatical and visiting another part of the globe.

All this on-line capability will allow students to play a much greater part in designing their own learning environment. The shift will be "from scheduled classes to individualised programs; from teacher-controlled to learner-controlled; from printed text to electronic materials; from memorising to problem solving and decision making". The orientation will move from content to competence.[56]

There will be a very different relationship between teacher and student: the teacher will be the facilitator, rather than the one who knows (as has been the case in the past with print). There could also be more genuine partnerships forged in electronic learning contexts.

For university teachers, some things will remain the same but much will be dramatically different. Those who are actually engaged in teaching will still have a considerable amount of preparation to undertake, but it won't be for a print handout or notes to put on the blackboard. Instead, lesson planning could take the form of software creation: individual programs tailored to the students' specific needs; the construction of "intelligent agents" designed to fit the idiosyncrasies of each student and to assist in navigating the complex, international, databases. In the way that the librarian was once on-hand to guide students through the catalogue and the various sources, a teaching agent could be available to help connect students to the Internet and the world information systems.

At the moment, this generally means helping them to access print-based databases. There's not a lot of video capacity on-line. (And if everyone wanted to use it, the entire cyber-system could come to a standstill.)[57] But the virtual-reality classrooms and simulations are almost here. For the student, this will be an enormously exciting experience; for the teacher, a performance challenge; and for the teaching agent, an educational support role.

Without leaving the computer terminal (at home, or at school), students in the cyber-classroom will be offered mind-boggling facilities. "Students can travel through on-line simulations to distant museums, laboratories or other cultures", says Parker Rossman,[58] outlining some of the possibilities. "Virtual-reality technology could create a 'history room', [for example] in which a scholar (or student) could enter a specific period of history or participate in a past event."[59]

Howard Rheingold, the expert on virtual reality (VR) states the obvious: that we are only on the brink of conceptualising VR potential. At the moment we are like

---

56. Edward Yarrish, 1991, "The Fully Electronic University", paper presented at Applications of Computer Conferencing Conference, Ohio State University, 13–15 June, quoted in Rossman, *The Emerging Worldwide Electronic University*, p. 99.

57. "Late Night Live", ABC Radio National, 5 September 1994.

58. Rossman, *The Emerging Worldwide Electronic University*, p. 96.

59. Rossman, *The Emerging Worldwide Electronic University*, p. 97.

people at the beginning of the century who tried to appreciate the possibilities of the film industry while they watched the first silent movies.[60]

But if there is only speculation about the specifics, there is widespread agreement among the experts about direction.

Perhaps one of the most important contributions that individual universities will make to the global learning facility will be the "teachstation". Such a platform will be required to handle the ins and outs of the educational environment. It will demand specialised software designed to help students find their way around the cyber-strata. (Mosaic from the University of Illinois is one example of this trend.) Perhaps it will be universities that specialise in the development of intelligent agents[61] that work for students to locate the information that they need from the mass of available sources.

Universities that produce intelligent agents for those who want to study music, dance, history or philosophy, for example, could well find themselves a niche market – and develop their own electronic identity in the process.

Music teachstations that allow composers to work and play together when geographically separated; platforms which allow students to practise along with an orchestra that is rehearsing in another country: these are the amazing possibilities that will be available and which universities can legitimately provide.

Luke T. Young et al.[62] have produced a sketch of a day in the life of a university teacher by the end of the century.

The teacher prepares the presentation, which of course includes sorting and sifting through a considerable quantity of graphics and electronic sources. It could take weeks, if not months, to get this into shape. And then on the day, perhaps, the presentation is made to a small group of students on campus, and simultaneously transmitted on-line to others all around the globe. (The presence of students, like a live studio audience, would provide reaction and feedback, so that the presenter didn't have to speak to camera.)

An electronic blackboard will be a vital teaching-aid for the presenter – who, it goes without saying, will have to be telegenic and a good communicator. The electronic blackboard will be viewable by all the on-line students, and it will be able to call up maps, charts, video, etc. As long as the technology works, the presentation of material could be every academic's dream.

The whole performance is recorded on video, so that it is available for future reference and for future students. It may even be assigned as a "text" in other classes, by other teachers in the department. It has the potential to be added to, updated, "cut

---

60. Howard Rheingold, 1986, "Virtual Communities", *Whole Earth Review*, Summer, and 1991, *Virtual Reality*, Secker & Warburg, London.
61. Intelligent agents are a software device which provides individualised service; like a butler, intelligent agents can search out particular requirements for their employer; they can organise material, get things ready, and generally serve the idiosyncratic needs of their owner.
62. Luke T. Young *et al.*, 1988, "Academic Computing in the Year 2000", *Academic Computing*, 2, 7, May–June, pp. 8–12, 62–5.

and pasted", both by the teacher who made it in the first place, and by subsequent teachers who use it for their own purposes, legitimately or illegitimately.

Parker Rossman then indicates how the teaching becomes the publication.

When it is ready then it may be "published", that is, made available for rent or sale for use anywhere with subtitles or translations into other languages. Before being thus published on interactive videodisk, it would be submitted to a peer review committee to see if it should be recommended for global use. For "the teleuniversity will enable the best teachers to lecture to thousands of students at dozens of universities simultaneously". However, the real advantage is that eminent researchers will be able to teach extremely specialised courses to small groups scattered round the globe.[63]

---

63. Rossman, *The Emerging Worldwide Electronic University*, p. 100, quoting from Young *et al.*

6

LIBRARIES

Public or Profit?

Lending and Linking

Last century, state libraries were established in the United Kingdom, the United States, Australia and other countries, with the express purpose of providing the public with information. On the grounds that everyone needed to have access to books, but that not everyone could afford to purchase them, the public library system assumed the responsibility of buying books for the community and distributing them.

The system has to be rated as a huge success. There are countless tales of scholars and specialists in all sorts of areas who acknowledge that it was the public library which gave them their start in life. There are millions of readers throughout the world whose major leisure activity has been associated with borrowing books from the library. Even in Australia, three times as many people visit a library as attend a football fracas.

Now that we are moving from books to terminals as the source of information, what should the role of the public library be?

On the grounds that everyone needs access to information, but that few can afford the expense of the latest technology, it could be argued that libraries should continue in their tradition. It could be argued that they should assume the responsibility for purchasing thousands upon thousands of terminals and providing them free to those who want to access the new information sources, in the same way as the community has been provided with free books in the past.

Will libraries allow electronic borrowing? Will the "library card" of the future entitle the user to VR kits, CD-ROM drives, special software and screens, etc.?

Already the State Library of New South Wales has moved to show the possibilities. Very soon, the library will be in a position to provide on-line services to its "clients", who will be able to access the Internet from their own homes. The only expense for users (those who have their own PCs with modems!) is the cost of a local telephone call to their local public library: this would probably still be much cheaper than any other form of "transport" to the information source. While users would have to pay for any commercial databases they might "surf" and download from, their link to the library Internet will be free in the same way that their link to books was, when print was primary.

Subtle and not so subtle shifts in this direction are already occurring. Partly because students were overloading the physical space resources of the NSW State Library, a Schools Research Service has been established. Any student, from anywhere in the State (and not just those who can make the trip to the library front door) can now request any information, and the library will provide.

The information is still in hard-copy form, delivered by snail mail or fax. But in no time at all – provided licensing arrangements are sorted out – all schools could be electronically linked to the State Library through the network of existing public libraries. Make a local phone call from any school in the State, and you can be connected to the international information network.

This is what Peter Lyman has referred to as the global reference room,[1] whereby all the libraries in the world are networked – to become one! The issue is, of course, to find your way around this vast resource. Librarians will become the information navigators – once they are adequately prepared for their new role.[2]

These new provisions raise some interesting questions in relation to government policies. It was partly because of the State Education Acts which insisted that all members of the community should be able to write and read that public libraries came into existence and made the necessary materials for reading available. It is not too much to predict that within a few years there could be Education Acts which make computer competence compulsory, with comparable implications for library services.

The recognition of the increasing gap between the information-rich and the information-poor has led to a growing appreciation that access to information – *for all* – needs to be enshrined as a human right. We have already begun to appreciate that literacy is a necessary condition of community participation, and "visualcy", or "informacy", is a logical follow-up. This is why the right to electronic information has been put forward as a proposition by Women, Information Technology and Scholarship (WITS) for the United Nations Conference on Women in Beijing in 1995.[3] If governments accept that *access* is a human right, what steps will they have to take to ensure that *all* members of society get on the electronic highway?

Of course there is the French option. Given the costs of providing every household with an annual telephone directory, and given that a high percentage of the entries were out of date by the time they came off the press, the French telephone company equipped

---

1. Peter Lyman, Follett Lecture, Internet, 4 October 1994, directed courtesy of Janine Schmidt and Colin Steele.
2. Interview with Alison Crooke, 21 September 1994; training of staff was one of the difficulties mentioned.
3. See "Gender Equity in Global Communication Networks: A Global Alert", Center for Advanced Study, University of Illinois, Urbana.

every household with a computer terminal.[4] France is but the first country in which the computer is becoming as much a standard household appliance as the television or telephone. (The trend is obvious in other countries. In 1994, "revenues for computers sold in the US were more than six times the revenues for TV sales, and home computer sales are growing ten times faster than TV sales".)[5]

Conscious of the growing disparity between the information-rich and information-poor, various libraries in the United States have taken initiatives to enfranchise more people in the local community. For example, Ken Dowlin, the librarian at San Francisco Public Library, has made explicit connections between information and politics. In his view, all clients (regardless of cultural or language background) should be provided with access to on-line information about social services and government policy. In this way, the community is politically empowered, and the library continues in the tradition of being a public information provider. According to Peter Lyman, this development is not unusual.

> Some of the most interesting visions of the future of the National Information Infrastructure are by social action groups, who envision the network as a dimension of government, not only for information about social services, but as an environment for political participation.[6]

Regardless of the solution, the problem is clear. There is a pressing need for a national (and international) policy on *access* to information, and the system of public libraries could well play a critical role in the development of such a network.

At the moment, without coordinated approaches, there are too many predictable duplications and gaps. Too many different government agencies at too many levels are trying to establish their own patch. In Australia, it could even be said that the issue of "States' rights" is rearing its ugly head as the various State libraries compete with one another, when arguably only *one* major resource is needed as a hub for the nation. It seems that the old railway gauge drama is being played out before us again, where every State goes its own way and the entire country ends up paying for such gross stupidity and lack of foresight. We cannot afford to make the same mistakes with the electronic highway.

Haven to Marketplace

It will come as no news to libraries – public, private or university – that they are at the centre of activity in the transition to the electronic format. In all the upheaval that is taking place, there are priorities to be decided: to stay with the book, or go with the computer? to preserve the past, or be pioneers of the future?

---

4. As Cheris Kramarae has pointed out, however, "In France, the national fiber optic Minitel sysem is largely funded by phone sex"; Cheris Kramarae, 1995, "A Backstage Critique of Virtual Reality", in Steven G. Jones, *Cybersociety*, Sage, Newbury Park, Calif., p. 46.
5. Donald Carli, 1994, "A Designer's Guide to the Internet", *Step-by-Step Graphics* (Peoria, Ill.), November–December, p. 27.
6. See Lyman, Follett Lecture, p. 1.

The directions that the various institutions will take are tantalisingly difficult to predict. It is not even possible to say whether, next century, libraries will see themselves as a public service or as an income-generating arm of the information industry.

> In one scenario libraries will act as navigators in the seven cyberseas of information, helping scholars surf monstrous flows of multimedia, full-text data pumped out from databases networked across the globe. In another, librarians will kill the book, sell themselves off to a private sector, and end up as pawns in the corporate battle for information control.[7]

In his article "Fear, Greed and the Electronic Library", Ian Reinecke (1993) paints a plausible picture of the privatised university library of the future. In his scenario, "The Vice Chancellor argues that outsourced management contracts are more efficient, that most of the university's service provision is unresponsive to customer demand and that privatisation will free up working capital and reduce the operating deficit."[8] There are seductive reasons for handing the library over to big business. This could be much more than a fanciful choice that university libraries will have to confront.

But there are no choices about whether the future will or will not be electronic; the revolution is already here to stay. There can be no going back to the libraries of the past, no matter how attractive they appear as retreats or sanctuaries. The resources and the users have changed irrevocably. Gone is the image of someone doing research or studying as a lone figure in a library, reading silently from a book. These days it is *activity* at a computer terminal that signals research and study – and stillness and silence, the hallmarks of book reading, are hard to come by.[9]

Alvin Kernan has deplored the passing of the old ways, and objects to the fact that libraries are no longer havens: "We shall not see the likes again of . . . such fictional readers as Mr Bennett who retreated to his library to escape the pressure of the household in *Pride and Prejudice*", he writes.[10] He goes on to lament that the privacy and quiet space that reading requires and that libraries have traditionally provided is increasingly difficult to find. Even the architecture gives cause for comment: "The buildings of the book, the great libraries and colleges begin to look quaint, like medieval cathedrals, and become inefficient and uneconomic."[11] Given the cost and time over-runs of the British Library, which is currently under construction, there would be widespread agreement today with this assessment of the "book-building" folly in the United Kingdom.

---

7. Eric Davis, 1992, "Cyberlibraries", *Lingua Franca*, February–March, p. 47.
8. See Ian Reineke, 1993, "Fear, Greed and the Electronic Library", in John Mulvaney & Colin Steele (eds), *Changes in Scholarly Communication Patterns*, Australian Academy of the Humanities, Occasional Paper 15, p. 163.
9. The young are now seen to suffer from Attention Deficit Disorder, characterised by hyperactivity with the mind and body racing around; but this could just be the consequences of an electronic society which demands rapid responses to the mass of information that confronts us. For further discussion, see Chapter 5.
10. Alvin Kernan, 1990, *The Death of Literature*, Yale University Press, pp. 134–5.
11. Kernan, *The Death of Literature*, p. 135.

Of course the writing has been on the wall for state and university libraries for many years. Print has produced too much information for libraries to handle by conventional means. No one could devise a satisfactory manual method of cataloguing, indexing or cross-referencing – or even shelving and storing – the 94,000 journal titles that are taken annually by the University of Illinois, for example.

Although there was once the cherished aim that the British Library and the Library of Congress would hold a copy of everything that has ever been printed in English, this policy has long been abandoned. Hard decisions are being made. The Library of Congress, for example, currently holds 88 million items; it receives 31,000 new books and journals every day, but it keeps only 7000 of them.

This practice of culling has implications for authors of every type of material. Academic writers are no exception. In universities where the premise has been publish or perish, the unthinkable is happening. University researchers and teachers are being encouraged to reduce their publications output. As Jennifer Kingson explains in her article "Where Information Is All, Pleas Arise for Less of It":

> the multiplicity of mediocre publications makes it impossible to sift out the ones that contain fresh ideas. The proliferation of books and journals seems to have narrowed access to information instead of widening it, and some universities [Harvard Medical School is her example] are considering ways to encourage their faculty members to publish less, not more.[12]

Clearly, something will have to be done to reduce the amount of print that is to be stored; it's too much, even with a radical selection process in place. The new British Library is a case in point. While it is to have eight miles of shelving, the estimate is that this will be sufficient for no more than ten years. And what then?

One of the problems is that each day more and more books are being published and more and more libraries are being built. This can't go on indefinitely. To Alvin Kernan, the only sensible solution is to store the print material in central databases – which calls for completely different architectural requirements to those of the reading room of a public library. "But such is the tyranny of custom," he comments, "that librarians go on building libraries and extensions that are out of date even before they are funded."[13]

Already, conventional libraries are well on the way to becoming electronic reference centres, like it or not, even when their priority is to preserve the culture of the book. Paradoxically, it is only through electronic means that many books can be saved for posterity.

The reason is not the deterioration of the pages, although this is threat enough: Eric Stange predicts that, as a result of the acid paper that has been in use since 1870, "At least 40% of the books in major research collections in the United States will soon be

---

12. Jennifer Kingson, 1989, *New York Times*, 9 July, p. E9; quoted in Kernan, *The Death of Literature*, p. 139.
13. Kernan, *The Death of Literature*, p. 139.

too fragile to handle."[14] The reason is that even now there's no such thing as "pure print". It is being produced and stored electronically, as Erik Davis illustrates:

> The information explosion may have ratcheted up the sheer number or printed texts, but the Book as an ideological, economic and scholarly entity is wheezing hard. New books, *qua* books, seem almost after the fact. Texts are generated on word processors, they're then sent via modem to a publisher's computers, prepared as digital objects, and only then printed. To be quoted, which would soon probably mean to be searched with keywords, they most likely will have to be returned to digital form.[15]

In the push to go electronic, however, an even greater pressure than that of space and preservation is likely to be economics.

> A panel of publishers reported that the book that used to cost $3 now costs $30 and will in the future cost $300. By that time the same information on disk which used to cost $300 and now costs $30 will then cost $3.[16]

That's not the end of it either, as far as Parker Rossman is concerned: "Furthermore, electronic books will not as often be discarded but can be updated from year to year for use by the next generation of students." This makes the electronic medium even cheaper.

Cutting and Merging

The pressure has been on libraries – particularly academic ones – for the past few decades, not the least because of the peculiarities of academic publishing. There was, for example, an explosion in the number of journals that appeared during the 1980s and it could be argued that this was the result of the motive of publisher profit rather than a flowering of the expansion of knowledge.

Robert Maxwell certainly exploited the academic conventions and the commitment to "keeping up with one's peers". He found that all he had to do was to set up new journals, and the academics themselves would take care of the copy and the sales. Free of charge. (Another part of his strategy was to sue anyone who tried to suggest that he was doing less than behaving in a beneficent manner; so much so that it would have been difficult for this "discussion" to take place when he was alive and litigious.)

Making use of the services of academics was honed by Robert Maxwell to a fine art. On the grounds that a specific publication was absolutely essential for the conduct of proper research, he would enlist the help of interested academics who would then put pressure on their libraries to purchase this "latest and greatest" on the market.

---

14. Eric Stange, 1987, "Millions of Books are Turning to Dust – Can They Be Saved?", *New York Times Book Review*, 29 March, p. 3.
15. Davis, "Cyberlibraries", p. 50.
16. Parker Rossman, 1992, *The Emerging Worldwide Electronic University*, Greenwood Publishing, Westport, Conn., p. 70.

While Pergamon Press customarily bore the costs of printing and distribution, it was academics who ran the journal, provided the copy, developed the "product lead" and placement, promoted the publication and pushed its sales. And all for nothing.

It was the librarians who started to ask questions and to protest at the voluminous expansion of scholarly publishing. Their concern was not that the system was based on the unpaid labour and exploitation of their colleagues. The librarians' grievance was self-interested; they began to claim that they were a captive market for unscrupulous publishers of academic material. They pointed out that the more journals that Pergamon, for example, produced, the more libraries had to pay.

This is when culling commenced in earnest. The contribution and costs of specific journals were scrutinised, and some questionable practices were exposed. For example, librarians found it galling that the library rates for Pergamon journals were considerably higher than the individual subscription rates. (In the war between libraries and publishers, there were more than a few libraries which resorted to subscribing to journals in the name of individuals in order to reduce the costs and the publishers' profits.)

Librarians also began to query the *relative* costs of journals; why were slim, scientific ones so much more expensive than bulky, humanities ones? It couldn't be that the contributors were paid more for their work. Given that it costs no more to print articles on stress in bridges than it does to print articles on stress in human beings (with relevant graphs), it was clear that the higher rate of the engineering journal had nothing to do with the cost of paper and ink.

But it had a lot to do with academic traditions, and the power, prestige and purses of disciplines. Those who could pay – would pay! Even then, the humanities and social sciences were the poor cousins who couldn't afford the full range of relevant publications.

At the same time that the cost of scholarly journals was soaring, libraries were confronted with dwindling budgets. This was partly because for the first time they faced competition when it came to information funding. Computer Information Services (generally managed by a male) began to appear on campuses and to attract resources which had once gone exclusively to libraries (an occupation where women had played a major role). The battle for money was on in more ways than one.

Throughout the 1990s there was considerable political activity on the part of libraries and their professional associations as they tried to deal both with rising costs and with increased competition from computer centres. In the USA, for example, associations of librarians were formed specifically to challenge what were seen as the monopolistic and mercenary machinations of some academic publishers, and the predatory practices of the techie-heads who wanted to take over.

Some of these battles have been lost and won. Some universities have now merged their libraries with their computer centres,[17] and many would argue that this has been

---

17. See Rossman, *The Emerging Worldwide Electronic University*, p. 62.

at the expense of the traditional ethos and functions of the library. Even where libraries continue to be an entity and to provide some of the old services, there is enormous pressure on them to change in the direction of the electronic medium.

The End of a Tradition

With the transformation of libraries comes the transformation of the role of librarians. The best that can be said of conditions in this traditional book-based occupation is that confusion reigns. "That librarians [in the US] are undergoing an identity crisis is perhaps best illustrated by the disarray of their training institutions", comments Erik Davis. "Since 1978, more than a dozen library schools, including those of Case Western and Chicago, have closed. Columbia University withdrew its support for the once renowned School of Library Sciences in 1990."[18]

Library studies are being discarded in favour of information science, but some of the old schools with particular philosophies and priorities have not "restructured" to meet the demands of the electronic age. To many librarians – steeped in the values of preservation – the end of these old schools and their practices is nothing other than a dreadful loss.

There are those who think that to abandon the traditions of the past in favour of the computer culture of the present is to "throw away the keys to the kingdom". But their protests and their resistance are relatively ineffectual.

Warnings about the loss of standards and the erosion of working conditions have not reduced the enthusiasm in some quarters for the gratification of information science. With the change in gender ratios among librarians has come the gung-ho adventurousness of many of the contemporary male pioneers who set off to conquer cyberspace. Calls for greater examination of the costs and the consequences – for a more cautious and critical approach – seem to have little impact.

> After this generation is through celebrating the unaccustomed freedom associated with data flow, the next is likely to look around and find itself enslaved in an electronic sweat shop, in thrall to data entry, production quotas, management supervision and the kind of drudgery generally associated with copy shops.[19]

The critics have a tough job, of course. It's very difficult to make a case against the delights and rewards of data flow. Even the most committed book-lover can be seduced by some of the marvellous things that the digital revolution makes possible. Because we are not just talking about access to catalogues, or even to the text of journals and scholarly publications. Once technological and intellectual property perturbations are sorted out, it will be possible to read everything that has ever been written, on your own screen.

The advantages of having the world's books on-line are obvious.

---

18. Davis, "Cyberlibraries", p. 48.
19. Davis, "Cyberlibraries", p. 48.

- You can search the "shelves" without ever having to stretch, climb, or get dirty.
- The entire library can keep "rearranging" itself, so that whatever you are looking for is always right there in front of you.
- You can move from subject to subject within the one book, or from one book to another, without having to do all the things in between, without turning even a single page.

Who wouldn't want such a facility? And this is just a small sector of the range of exciting services that the electronic library will provide.

Peter Lyman, a librarian of the old school who has made the transition to the new, doesn't like the image of the death of libraries and librarians. "New technology does not replace print", he argues; "It does things that print cannot do." But few seem convinced.

No one need doubt the intentions of the University of Southern California which "has just built a new Teaching Library designed to manage electronic and digital information" and which has the admirable aim of helping "faculty design information and to use it in their teaching and research, and to teach students how to access and to create new information sources".[20] But one can doubt that print and books and the traditional practices of libraries will be significant factors in the near future, once the new generation of cyber-navigators takes over.

Indexing

Putting Up the Signs

One way that cyber-navigators find their way around is by means of indexes and keywords. These are the road maps and signposts on the information superhighway, and they can either assist you in finding your destination, or make it virtually impossible for you to know where you are, or where you want to go.

It is a challenge to find the right keywords, the clearest classification system, which will help to direct the millions of people travelling through cyberspace. There are obstacles to be kept in mind. One of the first is that most of the directions are in English. Besides the fact that this is hard on the many for whom English is not the first language, there is the difficulty that a keyword or category in one culture may not translate well into the classification scheme of the English language. Possibilities for genuine misunderstandings and misdirections arise.

The language of communication in the sea and in the air is English, so it is at least consistent to have English (or, more accurately, American) as the language of cyberspace. But no matter how "sensible" the reasons, it is tough on the non-English-speaking world. There is a need for more research to be done on the implications of this – on international issues of access, power, and the development of the different "ethnic Englishes".

---

20. Lyman, Follett Lecture.

No doubt future generations will talk about the colonisation of cyberspace in much the same way as we look back on the nineteenth century and the imperial policies of the British. Just as we now think in terms of the British "colonising the minds" of many members of the Empire, so too will there be discussion of the way the Americans conquered cyberspace and became the cultural imperialists of the airwaves. The question arises – will we all think like Americans in the twenty-first century? (Some people feel that we already do; that the Americanisation of the world's cultures is already well under way.)

At one level, there's nothing wrong in thinking like an American. It's not a very productive exercise to try and rank one culture as better than another. But it is clear that the American language and classification scheme is sexist and racist – just for a start. This makes me very wary about the keywords and signposts that are being put up in cyberspace.

While they may prove to be a wonderful guide to white, middle-class, American-speaking males, the rest of us could find that they are not helpful directions for our purposes; it could even be that we don't exist and that there are no directions for us at all.

The sexist nature of the classification scheme of English/American has attracted feminist interest over the past few decades. Along with the irrefutable evidence that the English language has many more words for men – and many more of them are positive than are the words for women – there is the issue of who does the naming. Who has named the world, and from what perspective?

And with what consequences?

Not a few feminists have come to the conclusion that English is a man-made language.[21] Because men have held all the public offices until very recently, it is the names that men have given to the world which have become the official terms. Hence the greater number and the positive nature of the terms that relate to themselves.

Women might have generated just as many terms to name the world and themselves. But because women had no place in the political, educational, professional world, women's names rarely became the public ones which were known to the whole society and transmitted to the next generation.

Women and people of colour have had to find their way around in "white man's language" throughout the print period. They have experienced language as a stumbling-block. There are countless writers, black, female, who have written of the difficulties in trying to express their thoughts in terms that are not their own and which even as they write, denigrate and "second-sex" them. "This is the oppressor's language", wrote Adrienne Rich[22] as she attempted to move outside the established naming system to describe mothering, rape, violation from a woman's point of view.

---

21. See Dale Spender, 1980, *Man Made Language*, Pandora Press, London.
22. Adrienne Rich, 1971, "The Burning of Paper instead of Children", *The Will to Change*, Norton, New York, and quoted in Susan Brownmiller, 1977, *Against Our Will: Men, Women and Rape*, Penguin, Harmondsworth, Mddx, p. 334.

Now it looks as though the same limitations and biases that have existed in print will be transferred to the cyberworld. It's white middle-class, (Californian) men who predominate when it comes to choosing the categories, selecting the keywords, setting out the indexes, and putting up the signs. Those who are not white, middle-class, male, native speakers of English will be disadvantaged from the start.

As we leave behind print as the means of organising information, we are embarking on a huge project – to classify the world electronically. How we think, make sense, find our way around in the future, is going to depend to a great extent on the keywords, the indexes, the signs put up in cyberspace. It is tempting to name the indexers themselves as the makers of the new world. They will have literally the power of life or death over certain categories, experiences and ways of knowing.

If women aren't involved in the classification systems of the new electronic publishing, women will be excluded not only in the texts but also in the metatexts. As feminists have been arguing for the past two decades, culture is classification. Many librarians argue that a sexist database has been established for print sources: we need to pay immediate attention to this problem as electronic databases are established.[23]

Thought Management

When there is only a small group doing the naming – and white, middle-class males *are* a minority – it's easy for them to leave out the views and values of those who aren't members. Even the best-intentioned group that sincerely wanted to set up a gender-neutral sign system, for example, risks leaving out the experiences and understandings that are not known to it. And, of course not all such groups *are* well intentioned.

"Those who rule have the ruling ideas", and this is not good news for the ones who are ruled over. It means that we can be manipulated into "thinking" what the rulers want us to.

History is full of examples of particular individuals or groups who have colonised people's minds, who have made damaging use of propaganda. (Those who think that they can't be caught or conned by such uses are kidding themselves. We are all products of the naming schemes we are obliged to use, although some are better than others at identifying the blind spots.) If print was a powerful force in the wrong hands, the potential of the information revolution to get the population to think in a certain way, to adopt certain values that work against fairness and equality, is virtually incredible.

Thought management is a real possibility in cyberspace, and we should be thinking about ways to combat it.

Most of the material which illustrates the power of the index to mould people's minds relates to print sources. The implications can be transferred to the electronic environment and they show why there is cause for concern.

---

23. Maureen Ebben & Cheris Kramarae, 1993, "Women and Information Technologies: Creating a Cyberspace of Our Own", in H. Jeanie Taylor, Cheris Kramarae & Maureen Ebben (eds), *Women, Information Technology and Scholarship*, Board of Trustees, University of Illinois, p. 23.

It is not too much to claim that those people or events that aren't given a keyword or sign in cyberspace won't exist. In the print world, there have been all sorts of problems associated with the absence of a name: there is a classic feminist example that is relevant here.

When Betty Friedan undertook her groundbreaking and extensive research in the 1960s, she found that white, middle-class women in the United States independently described their shared depressing circumstances. But for Betty Friedan there was no existing category in which such data could be placed. She had to settle for "the problem without a name".

Interestingly, however, "the problem without a name" has itself become a name, a category, since her introduction of the term. It now stands for the many and various unnamed experiences of women. In any feminist classification scheme, dictionary or thesaurus, it would be entered as a keyword or category.[24] It's a term which originates with the experience of women – but it is probably not selected in any of the malestream indexing or keyword systems. And if women don't put it in, it won't get into the cyberspace indexes.

Another distressing example is where a category is quite deliberately omitted. Unfortunately, this is the case with the topic of violence against women in particular databases and research centres. There is more than one such place that I could name which has no entry for violence against women, and where "rape" is to be found under "life cycle". This constructs the entire area as "non-existent"; it implants the idea that rape doesn't exist, and that women who complain are "making a fuss about nothing". It also ensures that this way is barred to any who seek out information and who want to represent this idea in the community.

Yet another example is where the name used is so broad that it is impossible to find the precise material that you are looking for. Within the old print system, the Dewey classification system had so few categories for women that it was not unusual to find everything from whores to witches lumped together under the sign "woman". It was in response to this inaccurate, sexist, and negative naming that the first *Woman's Thesaurus* was devised.

> The language of standard indexing and classification system – terms used in most journals, libraries, filing systems, and databases – does not offer vocabulary consistently or sufficiently detailed and up-to-date to retrieve the wealth of [women's] resources available. Existing classifications frequently overlook emerging topics of special concern to women. As a result, important information is lumped together under "women" or is inaccessible.[25]

24. "Problem that has no name . . . makes women into cultural, social, and economic non-persons and robs them of their histories and their names": Cheris Kramarae & Paula Treichler, 1985, *A Feminist Dictionary*, Pandora, London, p. 358.
25. Mary Ellen Capek, 1987, *A Woman's Thesaurus; An Index of Language Used to Describe and Locate Information by and about Women*, Harper & Rowe, New York, p. viii.

Other areas of dispute revolve round issues of race and class. While the Library of Congress catalogue and the Dewey system of classification probably appeared neutral to those who helped create them – for whom they were a good fit – this is not how they appear to many people in society who have to use them, for whom the terms are not a good fit at all.

It is not necessary to be an expert in thought management to appreciate the way that certain terms can give a specific twist to the objects and events of the world. Sandy Berman is just one indexer who has challenged the signs that are going up, and who argues that the only roads that are being signed on the superhighway are those that reflect the habits of thinking of the oppressor.

> Item: *Jewish Question.* On the face of it, "Jewish Question" may seem like a bland neutral term. Yet it is just the opposite, masquerading ruthlessness and inhumanity – the age-old and altogether vicious practice of scapegoating – in a deceptive, leisurely abstraction. The phraseology is that of the oppressor, the ultimate murderer, not the victim.[26]

Sandy Berman is the plague of conventional librarians committed to providing a classification system that they believe describes the world as it is (and therefore determines how people will access it). He has cited countless examples of the way that the white, male, Western and Christian view of the world is represented in the categories.

Whose term is "Yellow Peril" he asks, for example? Surely this is not how the Chinese or Japanese experience themselves? And what about the colonialist mentality of the University of Zambia, which slavishly conforms to the respectable scholarly standards of the Library of Congress and therefore makes use of "Kaffirs", an offensive term for black South Africans?

Comparable examples can be cited in Australia, where offensive terms such as "half-caste" and "gin" are used and justified on the grounds that they have historical validity and that these are the subject headings that people are likely to seek information under. Needless to say, any offensive terms for white males have no place in the system; "genocide", "wife beating", and even "violence against women" as referred to above, are peculiarly absent despite their demonstrable and widespread existence.

Indexes and catalogues still contain such entries as "Discovery of Australia" by Captain Cook and "Discovery of America" by Columbus, showing the extent to which the white Western version of experience dominates. How do the Indigenous people of Australia and the United States feel about being discovered, I wonder? What were they doing before they were discovered? There's every suggestion, of course, in white Western terms, that they didn't exist before then. Such is the power of names to create the reality in defiance of the evidence.

---

26. Mark Pendergast, 1989, "Sandy Berman: A Man for All Subjects", *Wilson Library Bulletin*, March, p. 50.

It is clear that there are those who wish to hang on to the names and the power base they represent. The "invasion" of Australia, for example, is usually written up in the history books as "settlement" or "occupation"; recent attempts to change the name to "invasion" – which reflects more accurately the experience of the Indigenous people and conforms to the dictionary definition of the word – have met with fierce resistance in some quarters. There are many non-Aboriginal Australians who want to preserve their positive image; they do not want to have their antecedents cast in the role of invaders.

The list of examples of bias and distortion could be almost endless. Again and again we find that a particular name has misled, misdirected or deceived us. We need to remember as we move on to the superhighway that we are even more dependent on keywords in the electronic community. Energy should be directed towards ensuring that no misuse is made of the system.

To some extent, the rest of the world is going to be colonised by US terms and values. (There are good and bad things to be said about everyone watching the news from CNN.) But the extension of US cultural terms outside that country's borders is a different process from one where specific interest and power groups set out deliberately to manipulate the minds of millions. Sometimes they even claim that they are defending free speech. (There is more discussion of this area on p. 217.)

WOMEN, POWER AND CYBERSPACE

Been There, Done That!

Past Lessons

Looking back, we can see that in many ways, women were worse off after the print revolution than they had been in the manuscript era. (Women have often been worse off after a revolution. The French Revolution, for example, promised liberty and equality, but women were excluded from the fraternity and they lost out when it came to property and education.)

Given our history, it's not possible to assume that women will automatically share equally in any gains that come from the present information revolution. Women were excluded from the process of knowledge-making when the printing press was invented; and there's plenty of evidence today to suggest that women are again being kept out of the production of information as we move to the electronic networks.

An account of the way women have been prevented from being the knowledge-makers is eye-opening history, and should be taught to every one of today's students. But it is also a lesson in how, why, and at what cost to women and to society, females are set up as the second sex and as the subordinates.

No More Nunneries

In the post-print period, as today, women did not hold positions of power and influence over men in the Church. But during the Middle Ages there had been "women's places" where the Abbess or Mother Superior was in charge and where "women's values" could hold sway. Besides providing an alternative to marriage (and a refuge for women who wanted to be free from the bonds of family), these nunneries, abbeys and convents served as centres of women's traditional knowledge; skills in relation to plants, herbs, drugs and the natural world were often taught along with the sacred texts.

A few women, such as St Radegund of Poitiers, enjoyed a measure of fame and fortune throughout the scribal period. There was the German scholar Hildegard of

Bingen (1098–1178), who has been described as "the most notable medieval woman author on medicine, natural history, and cosmology".[1] The English contemplative nun Julian of Norwich (1342–c.1416) was held in high regard, as was the remarkable writer and philosopher Christine de Pizan, who, although not a nun, has been given the status of "France's first woman of letters".[2]

It would not do to overestimate the significance of all this, but it's clear that these women did experience a degree of recognition and affirmation during the manuscript era. But any power or prestige they possessed was drastically reduced once print appeared. For a whole range of reasons, women suddenly found themselves completely cut off from the new medium – and from any positions of influence.

The Reformation – a by-product of print and posters and protests – was just one of the forces which played a crucial role in diminishing the status of women.

For example, in the Protestant countries, women lost out enormously as the old religion was rejected, and the convents and abbeys were closed and their resources confiscated. This is precisely what happened in England when Henry VIII shut down all the religious orders and appropriated their holdings. Suddenly, almost overnight, the little education that had been available to women ceased to exist.

Of course, the monasteries were closed as well – but the consequences for men were not catastrophic. The end of the religious orders did not necessarily mean the end of their education. At the same time that the religious institutions were being shut down, the system of secular education – the university – was expanding. But only men were permitted entry to the new institutions. Women were not allowed.

When it came to learning, there was simply nowhere for women to go. As the historian Antonia Fraser says, this terrible blow to women was to have repercussions for centuries. "The disappearance of the convents at the time of the Reformation deprived English girls not only of convenient places of learning, but also of a pool of women teachers in the shape of the nuns themselves."[3]

Gone were the scholarly and often brilliant women who had made their religious communities centres of learning; gone were the models of women's wisdom and the traditions of enquiry and teaching. With them went many of the possibilities of girls' education for hundreds of years to come.

Men were able to take advantage of the printing press and the new forms of knowledge it made available. They could throw off some of the narrow teachings of the Church and begin to explore new methods of enquiry in the relatively free-thinking atmosphere of the university. But while men were establishing themselves in their new educational institutions, and laying the foundation of the scientific revolution, women were stopped at the door.

---

1. Londa Schiebinger, 1991, *The Mind Has No Sex?*, Harvard University Press, Cambridge, Mass., p. 13.
2. See Charity Willard, 1984, *Christine de Pizan*, Persea Books, New York, p. 15.
3. Antonia Fraser, 1984, *The Weaker Vessel: Women's Lot in Seventeenth-century England*, Weidenfeld & Nicholson, London, pp. 123–4.

This wholesale denial of education to women meant that they had to start from the beginning. From the sixteenth century onwards, women had to put their energy into *reclaiming* their educational rights. They were forced to prove that they were capable of being educated, and that they were "fit" candidates for university; needless to say, these were qualifications that were not required of men.

Throughout the nineteenth century, and well into the twentieth – and, some would say, even to the present day – women have had to concentrate their efforts on obtaining equal educational opportunity. It wasn't until 1948 that women were allowed to become full members of the University of Cambridge. But if the men were reluctant to accept the women themselves, they had no such reservations about taking their resources. Cambridge University was one place which benefited greatly from the closing of the convents: The "revenues and lands of the nunnery of St Radegund, an important centre of learning for women, were transferred to Jesus College Cambridge," comments Londa Schiebinger.[4]

Wise Women and Witches

Barring women from universities was not the only way that women's traditional knowledge was destroyed. It is no coincidence that men were setting up the new methods of enquiry which would form the basis of the scientific revolution and the contemporary professions, at the same time that women were being burnt as witches.

That this awful period of women's history is not generally researched or studied is interesting in itself. It shows that women have not had their own voices to describe, explain – and protest – throughout most of the print era. But because of the exclusion of women's experience from the history books, it's difficult to be definitive about the number of women who were boiled in oil or burnt at the stake after the invention of the printing press.

A historian of the scientific revolution, Brian Easlea, states that "more women were executed for witchcraft than for all the other crimes put together"[5] and he sets the number at tens to hundreds of thousands. The nineteenth-century feminist scholar Matilda Joslyn Gage puts the figure much higher. She establishes that more than 100,000 "witches" were put to death in France during the reign of Francis I. The Parliament of Toulouse, for example, burnt 400 witches in a single session. In her estimation, the total figure is closer to a staggering nine millions – and this was after 1484.[6]

Theoretically, women were witches when – like Eve! – they did deals with the devil, although the charge against them took many forms. Far from being the least-resourced members of society, witches were often the most beautiful, the most wealthy, and the most learned.

---

4. Schiebinger, *The Mind Has No Sex?*, p. 13.
5. Brian Easlea, 1980, *Witch-hunting, Magic and the New Philosophy: An Introduction to the Debates of the Scientific Revolution, 1450–1750*, Harvester Press, Sussex, p. 3.
6. Matilda Joslyn Gage (1893), 1980, *Women, Church and State: The Original Exposé of Male Collaboration against the Female Sex*, Persephone Press, Watertown, Mass., pp. 106–7.

Accused of witchcraft, women usually lost any rights to their property and, according to Matilda Joslyn Gage, it was for their lands and their money (not to mention their "bewitching" talents) that women were executed. But even more important than their wealth was their information. The word *witch*, notes Matilda Joslyn Gage, once meant a wise woman.

As men sought to establish their own methods and claims to exclusive professional status, they saw the traditional knowledge and skills of women as a threat, and attempted to eliminate them.

Much of women's knowledge and wisdom related to healing. Throughout the preceding centuries it had been women who predominated as midwives, surgeons and physicians. Prior to the professionalisation of medicine in the hands of men, it had been women who knew about birthing, who understood the properties of plants and herbs, who performed cures and found pain-killers. Women's record of success was often much better than that of the men who embarked on a medical career.

The eminent philosopher Francis Bacon (1561–1626) was one man who was able to give credit where it was due. He openly stated that "women are more happy many times in their cures than learned physicians":[7] but such acknowledgment of women's abilities was rare. According to some commentators, professional jealousy and the struggle for the control of information that went with it were among the driving forces behind the naming and persecution of witches.

"Male physicians not skilful enough to cure disease would deliberately swear there could be but one reason for their failure – the use of witchcraft against them," states Matilda Joslyn Gage categorically.[8] Brian Easlea agrees that the ignorance and incompetence of many a male physician were "explained away" by allegations of witchcraft.[9]

Being a knowledgeable woman was enough to attract suspicion and accusation, states Matilda Joslyn Gage.[10] With so many women put to death for being intellectually curious or competent, it's no wonder that they became wary of practising their craft and of passing on their information. It was only a matter of time before women's traditional knowledge was lost or suppressed.

The right to know about "physic" and to heal was transferred to the university. And "since it was all but impossible for female healers to enter university, there was virtually no way in which a woman could gain respectability" or credentials, writes Brian Easlea. To make sure that women's knowledge was disallowed and that only men had access to the new information, the male physicians "petitioned kings and parliaments to make the unqualified practice of medicine illegal. So the women healers, especially the more successful of them, necessarily found themselves on the wrong side of the law."[11]

---

7. See Easlea, *Witch-hunting, Magic and the New Philosophy*, p. 39.
8. Gage, *Women, Church and State*, p. 105.
9. Easlea, *Witch-hunting, Magic and the New Philosophy*, p. 39.
10. Gage, *Women, Church and State*, p. 103.
11. Easlea, *Witch-hunting, Magic and the New Philosophy*, p. 38.

Foolishness and Femininity

It can be stated without reservation that in the period after the introduction of print, the traditional knowledge of women was all but eliminated. Women who were seen to be powerful, who had information and could perform cures, were the ones who were tortured and executed. This persecution of women was an act of terrorism which had far-reaching consequences.

Women were frightened into hiding their intelligence, into denying their skills. They were coerced into pretending that females were foolish and that males were intellectually superior. This oppression of women by men continued for much of the time that men controlled the information medium. It was very hard for women to find any place of resistance. Not only were they prevented from practising and teaching their own knowledge, not only were they prevented from learning the new knowledge of men, but females were reduced to a state of dreadful ignorance at the time that men's knowledge of the world was dramatically expanding.

It is not widely known, for example, that in Britain, the education of girls was cut back after the introduction of the printing press. The "quality of public education offered to girls went downhill from the late sixteenth century onwards as the practice by which a few girls had attended the grammar schools, if not to an advanced age, ceased."[12]

As books became cheaper and more readily available, every effort was made by professional men to keep the new sources of information from getting into the wrong hands. They were clearly very worried about the fair sex; they didn't want women to be able to pick up books and read them – *and get ideas*.

Drastic measures were taken. In 1594 for example, Banbury Grammar School "forbade the inclusion of girls above the age of nine, or *when they could read English*."[13]

All the while that women's educational opportunities were being restricted, those of males were expanding at every level. "As the boys' grammar schools themselves improved with the progress of the century," comments Antonia Fraser, "the rift between male and female education grew into a chasm."[14]

As the gap widens between women and men in computer education and use, we can see parallels between the introduction of the printing press and the move into cyberspace.

The Gender Gap

The Figures on Men

The world of computers and their connections is increasingly the world of men: as more research is done in this new area and more findings are presented, the more damning

---

12. Fraser, *The Weaker Vessel*, p. 137.
13. Fraser, *The Weaker Vessel*, p. 137, my emphasis.
14. Fraser, *The Weaker Vessel*, p. 137.

is the evidence. Men have more computers, spend more time with them, and are the dominating presence in cyberspace: "Make no mistake about it, the Internet is male territory. Considering its roots are sunk deep in academia and the military-industrial complex, that's hardly surprising."[15]

What is surprising, however, is how quickly this power grab has taken place. Computers have not always been seen as the province of the male. On the contrary, when computers first made their appearance in the business world, it was considered perfectly appropriate to place them in female hands. After all, they had keyboards: and women were the ones with typing skills and keyboard experience.

And there are some men in the academic and business world who still see computers as "word processors" to be operated by females, with the sole purpose of providing professional and executive men with a clerical service. You would have to be ignorant, of course, to think that this is what computers these days are for. But this does help to explain why these men are invariably older (the generation gap again), and why they resist the idea of doing their own keyboarding.

It is interesting that some of them genuinely feel demeaned by the loss of the shorthand typist and the arrival on their desk of a computer terminal. (It's a bit like losing servants and having them replaced by the vacuum cleaner and the dishwasher.) But these men – who comprise a very, very small proportion of the privileged population – have no heirs in the next generation to follow in their footsteps. Younger men see the control of computers as power, not as an unfair demand to do their own work.

Men of the electronic media generation have moved swiftly to claim the territory. And just about every survey indicates that when the men have moved in, the women have moved out. In the UK for example, in 1978, 28 per cent of the students enrolled in computer science courses were women; by 1985–6 this figure had dropped to 13 per cent, and the trend continues.[16]

It's the same in colleges in the USA, writes Barbara Kantrowitz: "Fifteen years ago when computers were new in schools, they hadn't been defined as so exclusively male. But now girls have got the message. They are staying away."[17] The latest statistics tell the sorry story of women's increasing marginalisation as the medium gains in importance:

> In the United States in recent years, women have earned about half of all associate degrees in computer science, more than one-third of the bachelor's degrees and 37% and 13% masters and PhDs respectively according to *The Chronicle of Higher Education*. Yet, only 7% of American universities' computer science and engineering faculty are women. Of that a meagre 3% are tenured.[18]

---

15. Margie Wylie, 1995, "No Place for Women", *Digital Media*, 4, 8, January, p. 3.
16. Anne Cole, Tom Conlon, Sylvia Jackson & Dorothy Welch, 1994, "Information Technology and Gender: Problems and Proposals", *Gender and Education*, 6, 1, p. 78.
17. Barbara Kantrowitz, 1994, "Men, Women and Computers", *Newsweek*, 16 May, p. 51.
18. Wylie, "No Place for Women", p. 3.

The pattern in high schools is much the same. A few notable exceptions make it clear that that girls can do computing (take Methodist Ladies' College, Melbourne, described in Chapter 5, for example, where the entire school is computer competent and many students are enthusiastic whizzes). But generally, we find that girls are on the outer in the use of the medium. As early as 1985, American women educators could see the gap that was developing, and expressed their concern about the growing inequalities in computer expertise. In their book *The Neuter Computer*, the Women's Action Alliance wrote:

> most optional computer users are male. During school, the kids who dash to the computer room at lunchtime are usually boys. Many after-school computer clubs are nearly 100% male. Programming electives have overwhelmingly male enrollments. A young boy's parents are more likely to buy him a home computer than a young girl's parents are, and their fathers and brothers are more likely to use it than their mothers and sisters. A young boy's parents are also more likely to send him to computer camp in the summer, especially if the camp is expensive.[19]

Boys have more to do with computers than girls. The usual explanation given for boys' greater interest and competence is that they have far more opportunity to develop their skills. In the UK, for example, "six times as many boys as girls have a computer bought for them".[20]

The argument, then, is that boys are more interested because they have more opportunity, and that they have more opportunity because they are more interested. This doesn't get you very far; particularly if you are trying to find ways of making computer access more equitable. There are, however, other ways of describing the present problem, and developing some policy possibilities.

The present circumstances are a result of boys demanding access to computers and getting their own way. In any average co-ed school, this form of behaviour is witnessed as a daily occurrence. The boys take over the terminals. Even little boys. Nola Alloway of James Cook University, for example, has found that even three-year-old boys in pre-school insist that the computers are the boys' territory, and the girls are verbally and physically driven away.[21]

Despite the extent and significance of this problem, few systematic research projects have been undertaken in this area as yet. The findings that are available show which way things are heading. The Scottish Examination Board 1991 results are not atypical of Western societies, and help to reveal the extent to which girls feel unwelcome in computer classes:

19. Jo Shuchat Sanders & Antonia Stone, for the Women's Action Alliance, 1986, *The Neuter Computer: Computers for Girls and Boys*, Neal Schuman Publishers, New York, p. vii.
20. Cole, Conlon, Jackson & Welch, "Information Technology and Gender", p. 78.
21. Nola Alloway, 1995, *The Construction of Gender in Early Childhood*, Curriculum Corporation, Carlton, Vic., pp. 36–42.

for every one girl taking computing studies at the Standard Grade, there were two boys. At Higher Grade, boys outnumber girls four to one. Also during the 1980s, the growth of computing studies in Scottish schools paralleled a decline in the entrance figures for females into higher education computing courses.[22]

Of course, formal qualifications are no more a guide to computer competence than literacy figures were a guide to the number of bookworms. And because at this stage most high schools are not part of the information infrastructure, the courses and qualifications they are providing are often quite inappropriate as achievement measurements. More important than the enrolment in high school classes is the experience – positive or negative – that individuals are gaining out there in cyberspace. And this is where the discrepancy between the boys' involvement and that of the girls is starkly contrasted:

> In a recent survey of the world wide web, conducted by James Pitkow and Mimi Recker of the Georgia Institute of Technology, 4,700 people responded to questions about the use of the Internet: 56% were between the ages of 21 and 30; *94% were male*; 45% describe themselves as professionals; 22% as graduate students; 41% said they spend less than 5 hours a week exploring the Internet; 21% claim to spend more than 10 hours a week; 38% prefer searching in text mode; and 33% prefer searching in visual mode.[23]

If computer competence were an optional leisure skill, or just another means of collecting your mail, the gender gap might be merely a fascinating phenomenon. But there is nothing optional any longer about computer involvement. The electronic medium is the way we now make sense of the world, and this is why women have to be full members of the computer culture. Women have to take part in making and shaping that cyber-society, or else they risk becoming the outsiders: they will be the information–poor, as they were for so long after the introduction of print.

The history of the car provides a good parallel.

There are still some countries in the world where women are told that they *shall not* drive cars, and where this readily turns into the assertion that they *cannot* drive cars. It isn't so long since Western societies mocked women's driving capacity and expressed their contempt for women's competence in a range of jokes about "women drivers".

Over time, the attitude has changed. In Australia this is partly because of the introduction of the breathalyser and random breath tests. Once when couples went out on Saturday night, the men were almost always the drivers, while women occupied the passenger seats. The fear of losing a licence soon changed all this. Driving ceased to be a prestigious activity and became a service activity, and women assumed their place at the wheel.

---

22. Cole, Conlon, Jackson & Welch, "Information Technology and Gender", p. 77.
23. Donald Carli, 1994, "A Designer's Guide to the Internet", *Step by Step Graphics* (Peoira, Ill.), November–December, p. 27, my emphasis.

Women's status as safe drivers was bolstered by the figures of the insurance companies: the notion that women can't drive has pretty well disappeared, in the space of one lifetime and in my own generation.[24]

Women, then, have been accepted as drivers. But what sort of a gain have we made?

Certainly, it is better to be allowed to drive than to be prevented from doing so as in some Arab countries; and it is better to be acknowledged as competent than mocked and ridiculed. We could hope that the same attitudes will be extended to women and the computer. But even as women get their cars (and theirs is usually the "second car" – another feature of being the second sex) and take to the road, we have to face the fact that the machines we are using and the highways on which we are travelling have been designed by men, with little thought to women's needs or interests.

Had women ever contributed to the design of roads and vehicles, there is no doubt that the entire system would look very different. The priority would not necessarily have been to transport one man from the suburbs to the central business district; equal consideration might have been given to picking up the shopping and delivering children within the local community. The assumption would have been that as children are always around, it would be absurdly dangerous to think up and construct the sort of roads that we are currently stuck with. And think about the buses. No woman would have come up with a bus design that makes it virtually impossible for a person with child, plus shopping, to get on or off.

Women's acceptance as drivers came long after all the design work had been done. long after the patterns had been laid down. And there are consequences; no matter how many cars are sold to women, no matter how many buses and routes are put on to meet women's transport requirements, women are for ever restricted to working with a product that men designed to fit men's lifestyles and hobbies.

The highway system did not originate from the experience of women: they have to accommodate its needs rather than having it accommodate their needs. Unfortunately, much the same thing is happening with the construction of the superhighway.

> The question isn't whether women will be overlooked. They won't. As 53% of the American population and the arbiters of where a lot of the essential household income is spent, women are a pot of gold. Sure, women will be able to clerk and consume via the information superhighway: about that there is no doubt. The question is *how effectively will women be able to author, explore, build and motivate*? There's no doubt that information is power. The way things are going, women could be, once again, on the outside looking in.[25]

Construction work on the information infrastructure is only in the early stages but, as a few women commentators have noted, you can see the reality of women's

---

24. "On average, male car drivers are 46% more likely to be involved in a fatal car crash . . . Women drivers are less likely to be involved in a crash for speeding and/or drink driving" *but* "Women are more likely to suffer severe injuries because of their physical vulnerability and their tendency to drive smaller cars": *New Woman*, 1995, February, p. 35.

25. Wylie, "No Place for Women", p. 3, my emphasis.

second-class status being put into place all over again. And as long as women have less access to computers, as long as they are denied full membership of the cyberspace population, and a voice and a vote in the creation of the new culture, there is no hope that things will improve. Margie Wylie, editor of *Digital Media*, states that "far from offering a millennial new world of democracy and equal opportunity, the coming web of information systems could turn the clock back 50 years for women".[26] I think she is optimistic; it could take much longer unless women claim their place very soon .

At the moment there are many barriers to women's participation. One of the most obvious is that it costs money to purchase a computer, training and, for most people, time on the net. Because women have on average less money than men, they can be disadvantaged. When they can't get into this new medium, their disadvantage – and their lower financial rewards – are being compounded.

> In principle, electronic media are available to everybody. In practice, however, the reality may subvert open access. The Bay Area Women in Telecommunications (BAWIT) collective argues that because women are generally lower paid than men, economics may restrict access. Women may simply be able to afford less access than men. Economics may be a special factor for single or uncoupled women without a university or occupational net link . . . And as Airlie Hochschild argues in "Inside the Clockwork of a Male Career" (which is actually about women's careers) women's career paths tend to be delayed, which contributes to deferring participation in activities, such as learning computer skills that would facilitate net access.[27]

It's a case of the information-poor getting poorer.

To return to the analogy of the car: the solution is by no means as simple as that of ensuring that women have a vehicle to take them onto the superhighway. They have to want to venture out, they have to have some place to go and some person to visit. There are few calls for women to make in cyberspace while men take up most of the territory. As Maureen Ebben and Cheris Kramarae have commented in relation to academic women looking for a reason to log on: "Access is meaningless if a woman in the English department can only connect with the engineering bulletin boards or the men's movement newsgroups."[28]

Delia Browne, a senior law tutor and computer sex researcher at the University of Auckland, adds her own version of this: "We have to get more women on the net," she says; "I'm tired of talking to boys."[29]

The virtual reality is a long way removed from the ideal reality that was predicted when the information revolution began. There was no shortage of forecasts about how

---

26. Wylie, "No Place for Women", p. 3.
27. Theresa Conefrey, Internet communication re gender and the net, 22 April 1993.
28. Maureen Ebben & Cheris Kramarae, 1993, "Women and Information Technologies: Creating a Cyberspace of our Own", in H. Jeanie Taylor, Cheris Kramarae & Maureen Ebben (eds), *Women, Information Technology, and Scholarship*, Center for Advanced Studies, University of Illinois, p. 16.
29. "Cybersex – Censoring Fantasy", *Communication Update* (Communications Law Centre, UNSW), 108, April 1995, p. 6.

the medium was going to promote equality. "One of the greatest strengths of e-mail is its ability to break down socio-economic, racial and other traditional barriers to the sharing and production of knowledge"[30] is one of the many statements that expressed hopes for a better society. But this is not how things have developed. Quite the reverse.

There is ample evidence that women are being further and further marginalised – and this has grave implications for their future status and wealth. It is perfectly possible that in the next century, the verdict will be that women were worse off after the electronic revolution than they were before it, unless drastic changes are made very quickly to ensure access and equity for women in relation to the electronic medium.

Technology Turn-off

The design of the cyberspace environment in the twenty-first century will not only be crucial to our quality of life in general, it will be fundamental to the distribution of wealth and power. From the software to the hardware, from the interface to the infrastructure, decisions are now routinely being made which will affect that future; and they are decisions which serve the interests and values of some social groups far more than others.

This is why Donald Carli quotes the appeal of renowned graphic designer, William Hilson:

"Designers, particularly female designers, need to get directly involved in the experience of the NET . . . they need to get involved in large numbers if they intend ever to play a role in shaping the form or the content of the Metaverse in a significant way."[31]

This is a response, in part, to the growing body of research which suggests that some girls don't like computers. Many of them don't like what computers represent: and they don't like the way they are required to use them. So calling for more women designers whose insights and contributions could provide a remedy, and make the platform more female friendly, is a reasonable place to start.

That girls do have a different attitude to computers *per se* (as well as different ideas about their place and purpose), is a premise which is becoming more widely appreciated. Any business that is serious about designing infrastructure which matches people's ways of working (as distinct from forms which have to be imposed) would need to entice more women into its employ. But the recognition that the computer is not as attractive to women as it is to men gives rise to a general question – why is this?

Any explanation for the negative response to the computer on the part of girls has to be in the context of women's general relationship to science and technology. And it is not a good one. There is a vast amount of information which indicates that in the

---

30. Susan Herring, Deborah Johnson, & Tamra DiBenedetto, 1992, "Participation in Electronic Discourse in a Feminist Field", in Kira Hall, Mary Bucholz, & Birch Moonwomon (eds), *Proceedings of the Second Berkeley Women and Language Conference*, University of California, Berkeley.
31. Carli, "A Designer's Guide to the Internet", p. 27.

past women and technology have not got along very well. But this is not because of some simple anti-machine disposition on the part of women, as is sometimes suggested.

The reality is much more complex. Women have a lot to do with machines; they are often dependent on them, and they may feel very positive towards them. But the machines that are part of women's everyday reality are not usually regarded as "sexy technology"; indeed the washing machine, the microwave oven – even the typewriter – are sometimes not thought of as technology at all. But this apparent paradox, whereby women use machines but are held not to, can be readily explained by recourse to a fundamental feminist principle: that despite any evidence there may be of women's achievement, when women do it, it doesn't count.

As Autumn Stanley has said:

> It is no accident that food comes first in the traditional list of human priorities; food, clothing, shelter. Indeed, so obvious a need can be taken for granted, and the technology connected with it dismissed as non technology. Yet the females who learned how to gather, transport and make edible the staple foods of early human clans and tribes, were on the cutting edge of technology at that time. Moreover, without their inventions, the human animal would have had little more chance for survival and high civilization than its primate relatives.[32]

It's almost as if men, by definition, own the prestigious machines, so that anything they take up (including a keyboard) becomes prized and esteemed, to the point where it soon becomes seen as a masculine activity. This would help to account for the three- to five-year-old boys who insisted that girls could not do computers,[33] and the six-year-olds who argued that "girls could not play with computers because they needed to be tough".[34] It says something about the power of the mind to construct meanings that defy the evidence if we can produce a society that comes to believe that it now takes "balls" to master a keyboard.

The issues of what technology is and the way it becomes gendered are complex and fascinating, and deserve continued attention. But running through the literature on this subject there is the constantly recurring theme of the role played by socialisation: the sexes are taught specific technological relationships as part of their gender identity.

Sherry Turkle from Massachusetts Institute of Technology (MIT) is one of the pioneers in this area, and she expresses no surprise that the machines that are associated with the male fail to attract women. "There is the legacy of women's socialisation into relationships with technical objects," she says. And "for many of them [it is] best

---

32. Autumn Stanley, 1981, "Daughters of Isis; Daughters of Demeter", *Women's Studies International Quarterly*, 4, 3, p. 290; see also Autumn Stanley, 1992, "Do Mothers Invent? The Feminist Debate on the History of Technology", in Cheris Kramarae and Dale Spender (eds), *The Knowledge Explosion*, Teachers College Press, New York, pp. 459–72.

33. Alloway, *The Construction of Gender in Early Childhood*, pp. 36–42.

34. J. Benyon, 1990, "A School for Men; Learning Macho Values at School", paper given at conference of British Association for the Advancement of Science, Swansea University, 24 August.

summed up by the admonishment, 'Don't touch it, you'll get a shock.'"[35] No wonder women are "reticent" about approaching a computer terminal.

But she also points out that it isn't just that girls are being warned off. There are many other subtle factors at work. Girls are also encouraged to leave the men to the machines, and to put their energy into relating to people. Such counsel in their childhood can make them wary of forming any attachment to a computer system.

Girls who are exceptionally good with computers (as were the group that Sherry Turkle studied) were still cautious about the influence of the machine, and about the sort of person they would become if they were computer (or maths) oriented:

> In high school, Lisa saw young men around her turning to mathematics as a way to avoid people and describes herself as "turning off" her natural abilities in mathematics. "I didn't care if I was good at it. I wanted to work in worlds where languages had moods and connected you with people." And she saw some of these young men turning to computers as "imaginary friends". She decided to avoid them as well. "I didn't want imaginary friends in a machine . . ."[36]

Lisa was one of Sherry Turkle's sample of twenty-five Harvard and MIT women taking and succeeding in computer programming courses, but Lisa's familiarity with the computer and her high degree of achievement did not change her apprehension. Indeed, it could have been said to have reinforced it. Most of the young women surveyed made it clear that when they looked at the people who are drawn to computers they rejected them as role models – they didn't like what they could see.

Such a response fits the data on how women feel about science: it isn't that they can't do it, and it is not necessarily that they don't like the subject. What they turn away from is the image of the scientist or the computer hacker. It doesn't fit with their notions of themselves as women.

To many young women who want to work in ways that make a positive contribution to their community (and there are plenty of girls who express such sentiments when thinking about a possible career), the scientist and the techie-head are a turn-off. This is why, for example, most girls don't look to the world of the computer buff and don't imagine themselves flourishing in such an environment:

> There are few women hackers. Though hackers would deny that theirs is a macho culture, their preoccupation with winning and with subjecting oneself to increasingly violent tests makes their world peculiarly male in spirit.[37]

This is not what many women want. Hence their understandable computer reticence, as Sherry Turkle calls it.

---

35. Sherry Turkle, 1988, "Computational Reticence: Why Women Fear the Intimate Machine", in Cheris Kramarae (ed.), *Technology and Women's Voices*, Routledge & Kegan Paul, London, p. 41.
36. Turkle, "Computational Reticence", p. 44.
37. Turkle, "Computational Reticence", p. 45.

And of course the shock value is still there as well.

Besides disliking the look of hackers and their conquering world, and being reluctant to have an intimate relationship with a machine rather than a person, women fear that they might break something crucial. It's not only that the machine might hurt them, but that they might hurt the machine. They might crash a system, or cause a problem that will lead to considerable expense or inconvenience. (Again, this is in contrast to many men who, in the real world, have little or no regard for the mess they may leave behind as they tinker with their machines.)

Jessie, one of the Harvard and MIT computer women, speaks for many when she says:

> I am still teaching myself not to be afraid of screwing things up. I think that being a "hacker type" correlates with things like having played with explosives or taken apart things or climbed dangerously up trees and that type of thing as a child. It seems as though women are less willing to take things apart and risk breaking them, to try things when they don't know what they are doing and risk getting into trouble"[38]

There are other factors at work as well, which are sometimes difficult to articulate. Crucial to girls' socialisation, for example, is the notion of contingency planning; of thinking ahead so that when you go out, you will be able to get yourself safely home. What time will you be leaving, how will you make your way, what will you do if your friends aren't there, if your partner gets drunk, if you miss the bus, and how will you report in to announce that you are home in one piece? All these are the serious considerations of young women, *before* they set out. Many women choose not to go out (or their parents do not allow it) until they know every detail about how they will get back. This way of operating in the world is at odds with the unpredicability and so-called anarchy of cyberspace.

For boys, who are socialised to take risks, to seek adventure, to prove their courage and masculinity by setting off into the great unknown with no concern for the consequences (and no consideration for the "mess" they may leave in their wake), the information superhighway is an open invitation. It is no coincidence that it caters for men's sense of adventure and conquest: it was men who designed the system. It shouldn't come as a surprise to find that they are quite "at home" in the world they have created; while women are more likely to be uneasy in the unknown territory in which they are trying to find their way around.

Put any group of girls and boys together with a "new" machine, and the gender-stereotypical responses are clear and unequivocal. The general rule is that the boys want to play with it to see how it works; but the girls want to find out how it works before they will try to play.

Women want to know the rules; they are socialised to find out how things operate and what the consequences will be before they take any "leaps". As Sherry Turkle says, this is no way to approach the computer and its near relation, the video game:

---

38. Turkle, "Computational Reticence", pp. 48–9.

it is almost impossible to play a video game if you try to understand it first and play second. Girls are often perceived as preferring the "easier" video games. When I have looked more closely at what they really prefer, it is games where they can understand "the rules" before play begins.[39]

This doesn't mean that girls aren't good at using the computer; there are countless examples of women's expertise. But it does mean that as the medium is presently formatted it is more in tune with the disposition and training of boys. As the medium has been designed by men. It is logical to suggest that if women were in on shaping and making the environment, if women had an input that reflected their views and values and methods of operating, then this particular turn-off for women could soon be overcome.

The comments from women who are computer-expert[40] are interesting. Having got through the "computing" and "word processor" stage, they are thrilled with the potential of the computer for human communication. Sherry Turkle, and Nancy Kaplan and Eva Farrell[41] (who also studied girls who participate actively in networks and who purposefully sought out electronic spaces), found that women like "chatting" on their e-mail and bulletin boards; they thoroughly enjoy nattering on the net.

Some also emphasised the role that the Internet played in "publishing" their stories. My eight-year-old niece wanted to be "connected" so that she could send the stories that she wrote to her aunts: immediately! So that they could see them on their screens. And give an instant response (theoretically speaking). Many girls would identify with this.

In my own case, I have to confess to a complete conversion to the computer. As someone totally committed to writing and reading – to whom books were the meaning of life – there was a period when I could not see what relevance the computer had to my work and my leisure habits. I am constantly reminded of the stand I then took.

Recently, I was giving a lecture on the role of the computer in the university. I was saying that universities were in the information-making game and that anyone who was not fully computer-proficient would not have a job in the near future, when one young woman interrupted me.

"Only two years ago," she said accusingly, "you stood here and declared that you didn't know what all the computer fuss was about – that all they could do was let you 'cut and paste' a bit more quickly. And now you are saying that it is shape up – or ship out!"

My only defence was that it was not two – but three years since I had revealed my resistance to the new medium, I had had personal experience of the joys of the computer. I value it, not as a word processor (although it has marvellous advantages in

---

39. Turkle, "Computational Reticence", p. 49.
40. For further discussion of women's skills and achievements, see pp. 227–47.
41. Nancy Kaplan & Eva Farrell, 1994, "Weavers of Webs: A Portrait of Young Women on the Net", *Arachnet Electronic Journal on Virtual Culture*, 2, 3, July.

the preparation of a manuscript), but for its capacity to provide me with international, interactive communication.

Despite my socialisation and my reticence with other machines, I love my computer. (I acknowledge that it has its faults. Had I had anything to do with the design, the hardware would come in a range of fashion colours and I would have a purple PC.)[42] For me, the medium is a more like a combination of a good book (or a library of good books) and a telephone. I would not want to be without it and there is no going back.

I am excited about where my computer can take me; I am exhilarated at the prospect of what I can do with it, and the people and the places I can access around the world. I suppose it is a bit like a car, with its promise of mobility and travel. I am delighted to have such an opportunity – which is not to say that I am without criticism of the information infrastructure.

But I am not at all ambivalent about the wonders that the computer represents.

"Do you waste a lot of time on the net?" I was asked recently.

"Do you waste a lot of time – reading?" was my only response.

Teaching and Techie-heads

There's not much point in looking to educational research to find out what is happening with girls and computers, or to obtain an account of the growing gender gap. So little work has been done in the area that there is a resounding silence in the literature and databanks. This is yet more striking evidence of the generational identity of the teaching and learning professions. A significant number of educators in positions of power and privilege aren't computer-proficient – and they are not supporting or doing research on this subject.

Among the few studies that have been undertaken is one in the UK, by Anne Cole, Tom Conlon, Sylvia Jackson and Dorothy Welch (1994). They put together statistics which reveal women's decreased participation in computer culture, and they also tried to provide some possible explanations for this state of affairs. The starting point for their study is that, when it comes to computers in the classroom, males are dominant.

Males "dominate the classrooms and male priorities have shaped the subject's image", they declare, citing other studies that support their stand.[43]

There is a substantial body of research which indicates that males have dominated the co-ed classroom and continue to do so, despite remedial measures.[44] They can even do this when the area is considered to be one outside their general expertise (as in English, where efforts centre on encouraging them to express their feelings, and in the cooking labs where their contributions earn them inordinate praise and gratitude).

---

42. Olivetti clearly agrees: "The Milan industrial designers . . . have transformed the personal computer into a sleek, savvy and very sexy appliance . . . The dusky pink/red laptop . . . was apparently quite a hit in Italy, particularly with women." Trudi McIntosh, 1994, *The Australian*, 13 December, p. 20.
43. Cole, Conlon, Jackson & Welch, 1994, "Information Technology and Gender", p. 78.
44. See Dale Spender, 1990, *Invisible Women: The Schooling Scandal*, The Women's Press, London.

But the male appropriation of computer terminals is strikingly different from the male dominance that is evident in other classroom contexts. It has more in common with the playing field, where male students can aggressively take charge of the equipment and prevent the girls from participating. In computerised classrooms, it is not uncommon to see boys of all ages actively seizing the machines, and physically pushing the girls away. This violence can be accompanied by taunts about how girls can't "do" computers, and how screens and controls are a boys' game.[45]

Nola Alloway observed pre-school children and the use of a scarce resource, the computer; and her findings are highly illuminating. There can be no doubt about the aggression and the hostility that the boys were prepared to engage in to keep the computer for their own use:

A powerful elite of boys emerged who clearly understood how to physically manipulate a situation. Various instances were recorded of boys pushing, shoving, hitting, pinching and, on one occasion, biting. At the climax of one particular tussle, the boy who was operating the computer was knocked sideways off his chair on to the floor. As boys jockeyed for right of use, threats and counter-threats were recorded: "I'll twist your arm." "So? I'll twist yours!"[46]

Such determined behaviour on the part of the boys was not exercised exclusively against the girls, of course. A pecking order was established among the males, with some boys not able to gain reasonable access to "the scarce resource". (As Nola Alloway points out, one boy, Zane, had more difficulty in getting a turn than any of the girls. While there are recognisable gendered codes of conduct, this doesn't automatically mean that all boys do well and all girls get left out.)

While a few of the girls were prepared to take on the boys who used brute force to monopolise the computer (with Jessica and Hayley combining a couple of times to "eject a trenchant male user from the chair . . . when emotional manipulation failed") the customary response of the girls was to stay away:

Girls, in general, seemed uninterested in accessing the resource when competing for the opportunity meant entering the fray with an antagonistic group of boys and the adoption of a combative style of interaction. Within this context of aggressive competition, it seemed understandable, if not outright intelligent, to avoid the computer and to select an alternative activity that did not involve physical harassment.[47]

Pre-school is only the start. So widespread has this male habit of taking over become, that I know of no co-ed high school – in Australia, the United States, or the United Kingdom – that has not had to come up with a policy which aims at ensuring some access for girls. Most schools have some form of girl-only time scheduled, and

---

45. There is more discussion along these lines on pp. 202–12.
46. Alloway, *The Construction of Gender in Early Childhood*, pp. 38–9.
47. Alloway, *The Construction of Gender in Early Childhood*, p. 40.

while this is far from an ideal solution (because it adds to the impression that girls need special assistance to do what the boys take for granted), at least it is an improvement on the denial of all access to computer terminals.

Nola Alloway also believes that there are problems with the strategy of girl-only time on the computer; while she is adamant that this is when girls do get their share of the resource, the arrangement fails "to address the asymmetries of power relations between the girls and the boys; the basis of the bargain that ensured that these pre-school boys dominated the computer and that girls' interests were subordinated remained undisputed. In brief, the power base was left intact."[48] But what else is there to do? "If we just allowed the students to use computers as they want to, girls wouldn't get a look in," one Australian teacher told me, summing up a general response.

In more than one school that I have visited, there were sufficient computers in the room for everyone to be working at their own terminal. But the girls could be found huddled in a group away from the computers, while the boys sometimes kept two terminals going!

In one such classroom I asked the teacher – whose manner was visibly hostile to the girls – why it was that the female students stayed together and made no use of the computers. I was informed that this was the girls' choice. That there was only trouble when the girls got to a keyboard. And that the reason they wouldn't "learn" was because they were out of their depth, and didn't want to appear stupid.

There could have been some substance in this appraisal – that the girls didn't want to look stupid, for example. But I thought that the statement said more about the teacher than the girls. There was no doubt in my mind that he would pounce on any "errors" that the girls might make, and that he would not hesitate to hold them all up to ridicule. But I didn't take him at his word; I checked with the girls.

"Why don't you work at the computers?" I asked them. Their answer was as one. "Because the boys won't let us!" they said. "And the teacher thinks that's OK."

They told me that they had made a fuss, that they had complained, that they did realise they were "missing out". But they also said that there was too much "aggro" in trying to change things – and anyway, they liked chatting to each other in computer class instead.

Wanting to find out more, I said that I found this hard to believe. (Neither the boys nor the teacher were party to this conversation.) If I were to visit the classroom again, would they let me see what happened when they tried to have a turn at the computers, I asked them?

They agreed – with considerable enthusiasm, I must acknowledge. The result was exactly as they predicted. Some boys physically pushed them away from the computers and insisted that they needed *two* terminals for the purposes of their project. They verbally abused the girls ("slut" and "slag" being among the most printable) and generally engaged in loud and bullying behaviour.

---

48. Alloway, *The Construction of Gender in Early Childhood*, p. 37.

Far from being embarrassed by this appalling display, the teacher felt vindicated. "I told you girls were trouble," he said. "They stop everyone from working."

I was stunned (and lamented the fact that I had no video camera with me). It took me days to recover. I still don't know how you tackle such a situation, short of calling for the dismissal of the teacher and for a course in decent behaviour for the male students. How many classrooms are like this, I wondered?

Obviously, teachers help to set the classroom climate in which such actions take place. I am convinced that some of these fifteen-year-old boys would not have behaved in such a mulish and macho manner had they not received the covert support of the staff. Indeed, I am sure that the parents of some of these boys would have been horrified by such goings on, and that some of the boys themselves would have been genuinely contrite in another context. But I do not think that will stop them from continuing their harassing tactics while ever conditions permit.

I could have counselled the girls to make even more fuss; I could have contacted their mothers and reported the abuse; I could have ensured that the fathers knew that their daughters were not getting a fair educational deal. But it can be irresponsible to stir things up when you can't stay round to support the people who will then be in the front line. So I said very little at the time: now I am saying a lot.

It may be argued that only a few teachers so blatantly structure the boys to be the information-privileged of the future, but I don't believe there are only a few bad apples: I think the attitude is much more pervasive. In addition, the majority of teachers, even those who are well intentioned, can still create a classroom climate which discriminates against the girls. In the UK for example, it has been found that teachers don't always subscribe to the idea that both sexes have the same need to be computer proficient.

Spear's research showed that 49% of teachers sampled (across all subjects) believed that technology subjects are very important to a boy's general education compared with 24% rating them as very important for a girl's general education. When asked about the importance of technical subjects for pupils' future, the gap widened further to 60% and 25.7% respectively.[49]

That teachers have differing expectations of girls and boys with computers is quite easy to demonstrate. In 1991 Sue Willis undertook some research on girls, boys and computer use – and discovered that there are different patterns of behaviour.[50] Her findings suggested that while girls worked through the computer exercises (which I would object to, to start with!) the boys were more likely to play games. And at the teachers' conference where these results were presented, there was general consensus that:

---

49. Cole, Conlon, Jackson & Welch, "Information Technology and Gender", p. 80.
50. Sue Willis, 1991, "Girls and Their Future", keynote address, national conference of Australian Women's Education Coalition, Perth, 7–9 June.

- boys played games because they were confident with computers, while
- girls worked through the instructions because they lacked confidence and were reduced to just following the program set out for them.

I wasn't satisfied with this teacher-interpretation, so I did my own little experiment. At a teachers' conference on the other side of the continent a little later in the year, I talked about this research – but I reversed the findings.

I informed the audience that it had been established that while girls played games, the boys worked through the programs. I asked them what they thought this meant. There was almost unanimous agreement; but while my findings were reversed, the teachers' profile of girls and boys and computer skills stayed exactly the same. They agreed that:

- boys worked through the programs, because they were confident and knew what they were doing, while
- girls played games because they didn't know about computers, lacked confidence and wouldn't tackle the serious stuff.

Now this is pretty depressing stuff. It suggests that *no matter what girls do*, no matter how well they achieve, it will be interpreted to mean that girls lack confidence. And no matter what boys do, no matter how poorly they perform, it will not interfere with the premise that boys are good with computers and therefore entitled to be confident.

It is another case of when girls do it, it doesn't count. And it has disturbing consequences.

A parallel can be drawn with the construction of maths anxiety which has happened to girls in the past. Because society believed that girls couldn't do mathematics, girls who did well were seen to have done something unusual. The girls themselves – who are also part of the society and share the belief system that girls can't do maths – soon started to think that they were lucky when they got a good result. And that next time their luck would give out, and they would probably hit the "right" level – at the bottom.

And boys? Well they had been reared with the notion that boys were better at maths. Why should they not believe it – and act confidently? And if they didn't do well, it was because they were unlucky, that their teacher was no good – or because they had been a bit slack during the week – but it certainly wasn't because the task was beyond them.

So we have created an educational culture in which girls who do well doubt their ability, while the boys who do poorly have no doubts about their potential to perform at the highest levels. We have artificially created anxiety among women students. We have handicapped them with our expectations. And we are on the way to recreating this pattern of significant inequality all over again – but with computer-performance this time round.

It is no coincidence that the computer is becoming a significant site of inequality, for it is at the centre of the future web of wealth and power.

There is no easy solution, either. There is a deeply entrenched belief that there's something wrong with women when it comes to computers (which is quite different from suggesting that there is something wrong with computers when it comes to women). Countering it is not as simple as giving encouragement to girls. After one classroom observation where I had witnessed girls being praised for doing things on the computer that had not been commented on when done by boys, the girls expressed considerable resentment.

"Doesn't she think we can do the work?" they muttered to each other, as the teacher lavished praise upon them and ignored the comparable achievements of the boys. "And why does she ask us all the dead-easy questions that a dork would know – and leave all the serious ones for the boys to answer?"

It's hard to win. The teacher thought she was doing the right thing in encouraging the girls; she didn't want them ever to feel dumb or silly, so she only asked them questions she felt sure they could answer. And it has to be said that they didn't appreciate her well-intentioned efforts. Nor did her attempts at support help to boost the girls' confidence.

In the repertoire of educational strategies it is difficult to know what to suggest to overcome this problem of actually creating gender disadvantage in the schools. There are not a lot of success stories of the past to draw upon. For example, we know that boys had greater difficulty with reading, and that huge resources went into trying to overcome their problem. But this wasn't necessarily a solution. As discussed in Chapter 3, a remedial reading industry emerged; the more kids there were with reading problems, the better the industry was served. So the numbers have steadily increased rather than decreased over the years.

That's not the only criticism of remedial reading programs. The fact that so much more money was put into the education of boys rather than girls is in itself is another form of gender bias. No equivalent resources were ever made available to encourage the development of mathematical skills in girls. Yet again, the double standard operates.

When boys had a problem with reading, every attempt was made to improve their performance; when girls had a problem with maths, it was a reason for advising them against doing it.

This is not a good tradition from which to start addressing the problem of gender inequities with the computer. Given the record of success of the remedial reading industry, I wouldn't want to argue for a vast number of remedial computer classes. Though it has to be said that the resources are needed to combat the sexism and sexual harassment routinely experienced by girls in the classroom. Such preventive strategies are certainly going to be a priority of the future.

The high school boys who hassle the girls and keep them out of computer classes are not going to be challenged to change their colours when they arrive at university. Many will soon find that the university computer science courses are even more supportive of abuse and harassment of women.

Entering the computer lab at university is for many female students just about the same as entering the men's changing-room at a major sporting event. The atmosphere is laden with testosterone, and the overwhelming message is that women are not welcome.

Lynda Davies is one of that rare breed of women, a professor of computer science at an Australian university; and she has been a whistle-blower in drawing attention to the rampant and ravening practices that characterise certain computer labs in Australia and other countries. The details are hard to take.

> During the day, computer labs are filled with students shouting across the rooms at each other, goading each other with terms like "fuckwit", "wanker", "dickfor" being called out as terms of comradeship, seeming terms of endearment amongst the peer group. The student groups are almost exclusively male, with females tending to be either quiet and contemplative, or chatty and smiling. The females do not use the same terms of comradeship for each other or for the males. They tend to use their personal names.[51]

Computer science labs across the nation – and in many other countries of the Western world – are not pleasant places for women to be. Lynda Davies says that the scene she describes is "normal" for computer classrooms; it is the rule rather than the exception to find that the labs are highly threatening and abusive environments for females. There is no mystery whatsoever as to why women prefer to be somewhere else, and to do some other course.

The male-dominance dynamics of the computer lab are no mystery either. This is turf-grabbing behaviour that the men are engaged in; by adopting such a macho attitude they not only advertise what they see as their masculinity, they also scare off the women and keep for themselves the power and the territory.

Just as the schoolteachers set the scene for the abusive behaviour in the high school computer room, the computer science staff can be responsible for the intimidatory atmosphere and antics of the labs in the university. In the case study provided by Lynda Davies, eight out of the nine staff members serving the computer labs were male; and only males supervised the students. The behaviour of many of them was worse than that of the student boys. And, she adds, "Female students have no choice but to conduct their work in these labs."[52]

Lynda Davies goes on to state that it is not unusual to find the male staff playing "games" in the lab (although officially this is not allowed): nor is it unusual to find them engaging in sexually explicit "virtual" activities even outside university hours.

> Late on a Saturday night, the technical staff (engrossed in arousing activities) can be found at the other end of the computer lab from the student engaging in the explicit delights of oral sex with a disembodied "female" partner by e-mail interaction. [All] of these computer users are male [and they] develop traditional male macho self images as they live out their fantasies by interacting with the screens.

---

51. Lynda Davies, 1994, "The Gendered Language of Technology", paper given at conference of International Communication Association, Sydney, 11–15 July.
52. Davies, "The Gendered Language of Technology".

This scene of violent, abusive male behaviour and brutally sexist language and imagery is horrific. It leads Lynda Davies to ask: why so extreme? What is it about computers and computer science labs that brings out this gross misconduct among so many of the men who work in the area: how could this be part of the academic environment of the technologist in training?

> It is not obvious that computer technologists are going to be gladiators needed to defend nations through brutal battles in which their lives are at stake so why should that hero image be so prevalent, and the aggressive, sexually dominant military image of the male be so dominating?[53]

What some of these men do on-line – what they put on their screens and on those of the women students – has to be seen to be believed.[54] To my mind it is nothing short of sexual terrorism, designed to drive women away from the centre of power. Last century there were laws against women being educated, against entry to the professions, against freedom of movement, against the right to be a person, to control your own fertility and to own property; in the circumstances, it was relatively easy to keep women in their place. But as we approach the twenty-first century, women have been successful in overthrowing many of the laws which discriminated against them – so it looks as though brute force and intimidation are being used to keep women out.

The parallels with the persecution of women as witches after the invention of the printing press could also be noted.

"Turf-grabbing", "territory-claiming", "boundary-marking" and the control of the centre of power; these are the most frequent explanations that have been given for the macho behaviour of the nerds and the techie-heads. But there's also another factor which should be taken into account. It's the masculine relationship to the machine, which seems to bring out the worst in some men. It's been there with cars (the biggest, brightest, latest, fastest) and it's there with computers as well.

Men tend to see machines as an extension of their own anatomy; where technology and gender intersect.

> The linkage of hardware to the penis is often made in jokes shared by computing students. This is witnessed in face-to-face interactions. "Ah, my machine's got more megabyte extensions than your machine" (thrusting of the pelvis and hand placed upon the groin); and in electronic interactions over the network. "What do you call a supercomputer? A memory with balls."[55]

This computer-penis mentality is not only found among the students in the computer lab: it's found in the support and teaching staff as well. It has serious implications for women who want to learn computer skills and who are confronted by the techie-heads who have trouble distinguishing their own parts from those of the computer.

---

53. Davies, "The Gendered Language of Technology".
54. There is more about this on pp. 202–27.
55. Davies, "The Gendered Language of Technology".

In universities where surveys have been done, the indications are clear. The computer support staff (often the graduates of the notorious computer labs) are almost all male. Although these people are charged with the delicate task of instructing faculty and students in the art of the computer, their teaching style is much the same as their user style: sexist and intimidatory.[56]

Many women on campus report that individual contact with support staff can begin with a man (or men) swaggering into their office. On seeing the model of the computer on the desk, there is the almost mandatory response, "Where did you get that? Has to be out of the Ark. You can't give that model away now."

This can be followed with a discussion on how much things have changed. If more than one person has come to help sort out the system, there can be an exchange (in which the woman cannot participate) on the latest whiz-bangery that is available, and which would be problem-free if only it were on the desk in front of them. So much of what is said is about measurement. How big, how fast, how many bytes and RAMs it packs. This hot-rod computer mentality again serves to remind the woman that it's a boy's toy that they are talking about.

Rarely is the computer support person like the thoughtful young man who comes to get rid of any bugs in my system. He respects my work, and sees his role as helping me to be able to do it with the aid of the computer. He makes no attempt to show me up as ignorant but designs programs that make it easier for me to do what I want to.

He stands while I stay seated at the keyboard, and he carefully talks me through the steps until I am confident that I can do it myself when he leaves.

From accounts I have heard, this is an unusual experience. Much more common is the computer support person who casts the client aside and after a series of flamboyant clicks, which emphasise his expertise and mystique, declares that all has been fixed, it was merely a matter of doing "X Y Z". He proceeds to depart – looking bored and disdainful.

The woman who has asked for support and instruction is none the wiser. It would be understandable if she were less confident. If she were left feeling frustrated, stupid – and that she had been a bit of a nuisance bothering the big man.

This is not how it should be. Support staff are not hired to impress with their technological prowess. Their role is to support and instruct – but very few of them do it. At the moment it could be said that some of them are simply adding to the problem of computer anxiety. They are reinforcing the stereotypes and the inequalities: they are making it difficult for women to feel at ease with the new technologies.

Currently, the area is controlled by the techie-heads; they are the ones who are calling the tune, and they do so in a way that protects and extends their influence. We need

---

56. WITS (Women, Information Technology and Scholarship) is a colloquium set up in the Center for Advanced Studies at University of Illinois, Urbana Champaign, with the specific aim of monitoring the impact of the new technologies on women on campus; seminars, lectures, publications, etc. have been directed towards documenting and analysing the experience of women – including that of their relationship with support staff.

good teachers to become computer support people; we need the arts and humanities values brought into the computer system. It might then be possible to change some of the blatantly sexist practices.

If we don't, then the only certainty is that girls face a bleak future. As society goes further down the computer path each year, girls will be left further behind, with no ready means to catch up. There is a bitter reminder here about the fate of women after the introduction of print.

After five hundred years, women were just beginning to look as though they were drawing even with the men. They have reached the stage in countries like Australia where, for the first time, more women than men have been gaining higher education qualifications. But this success has been achieved in an education system still based on print, where the skills needed to excel have been reading, writing and memory – all things that women are good at.

And just when it looks as though equity is about to be realised – the rules of the game are changed. The society (and soon, the education system) switches to the electronic medium. And "everyone" knows that girls are not as good as boys – with machines!

Girls can't afford to accept this situation. Demands have to be made. There has to be a campaign to change the climate of computing, and the nature and quality of support services which help women make the transition to the electronic medium. We have to come up with different hardware, software and personnel, where the focus is on women's way of knowing and doing.

In Britain, an MP, Judith Church, is trying to do something about the way women are being ignored by – or are ignoring – the computer industry. She has tabled a Bill to stimulate female participation in high technology.

> According to *Gathering Storm*, a report compiled by the UK Institute of Data Professing Management, women last year made up only 4% of IT managers, compared to 9% of all other management . . .
>
> The more technical the job, the fewer the females . . .
>
> This is reflected by the fact that girls account for only 17% of entrants to computer science courses today, compared with 22% in 1980, and that the percentage of girls studying A level computer science dropped from 22% in 1978 to 11% in 1993.
>
> Yet employers look for degree qualifications.[57]

The Bill focuses on the need for schools and employers to encourage women to be interested in computing; and government funding would be available for research on how to get women in – and how to keep them there. This is probably one of the first political/legal efforts that has been made to ensure that women have a share in the electronic future.

In the interim, if all else fails, there is one strategy to be kept in mind in this frustrating, transitional period: "I don't care what sex my next partner is," a young friend informed me. "Just as long as he or she can take the bugs out of my computer."

---

57. Elspeth Wales, 1995, "MPs Attempt to 'Feminise' IT as Participation Rate Declines", *The Australian*, 4 April, p. 27.

Toys for the Boys

Some people have been quick to point out that when it comes to the gender gap with computers, the fault may not lie with the girls; the reason for their non-participation could have something to do with the materials that are on offer.

Having spent some time looking at CD-ROMs and the computer games that are available – and these are the means by which many young people make their way into the cyberworld – I am of the definite opinion that any girl who refuses to play is showing very sound judgement.

Most of these games (and the video games which are really just another version of the computer game) are of the "drop-dead" variety. The main aim seems to be to kill as many people as possible, with a preference for violence against women. This goes quite a way towards explaining why women might not be attracted to such leisure pursuits. (It also raises some questions, however, as to why boys in general should find violence and the assault of women so engaging.)

One thing that becomes obvious in looking at the content of these computer games is that they are strikingly different from what is on offer to girls in books. When the standard narratives of girls' books are placed alongside the standard theme of computer games, the contrast is extraordinary. It is easy to see why girls go for the books rather than the software.

Girls are interested in the personal relationships, in the ongoing story of existence. Death in a novel could in itself be traumatising. Even now I can recall the terrible effect that the demise of Judy in *Seven Little Australians* had on me, along with that of Beth in *Little Women*. Two deaths – in the thousands upon thousands of stories that I read as an adolescent and a young woman – and they still have the power to distress me. I dread to think what would have happened had I been exposed to all those killer conquest computer games in which my sex was the target. Either I would have become very disturbed, or I would have been forced to change my value system. And that would have made me a very different person. Maybe even "more like a man". It's an issue worth thinking about.

To some extent we not only read what we like, but we are what we read. All the girls who are reading all the books about individuals, about relationships, about life patterns, are becoming more attuned to the role of the personal and interconnections in communities. They are being socialised as girls. They wouldn't learn these same lessons if they were reared on the diet of depersonalised characters and annihilation that are the ordinary stuff of computer games.

At the moment the girls are staying with the personal and away from the games. While this may be good in terms of their socialisation, it has been argued that it is bad for their entry to the electronic world. According to Professor Evelyn Rozanski, it's because girls don't play computer games that their participation rates in computer science have dropped from 40 per cent in the 1980s to less than 10 per cent in the 1990s. "If you are trying to interest a girl in computer technology, don't ask her to play a computer game," she says.[58]

---

58. Smith, 1993, Internet communication re July 1993 *Technology Review*.

Women who studied computer science in the 1980s were usually first exposed to computers at a time when the few games that existed like Pong and Pac Man were fairly generic and appealing to both girls and boys. Conversely, most women attending college in the 1990s were inundated during childhood and adolescence with male-oriented computer games that promote violence and stereotypical macho values. Girls not only get turned off to these games, says Rozanski, they turned away from computers and ultimately, careers in computer science.[59]

There is mounting evidence from the manufacturers of the games that their audience is almost exclusively boys. Joseph Pereira, writing in the *Wall Street Journal* says:

> Boys play video games far more than girls do. According to NPD Group which tracks retail sales, 75% of video games are bought for boys. The gap is wider for more advanced systems. A usage survey by Nintendo Ltd of its older games which run on an eight-bit chip, showed males outnumbered females 59% to 41%. Among users of its more powerful 16-bit Super Nintendo Entertainment System, though, 82% are male. Similarly, 80% of the players on Sega Enterprises' 16-bit-system are male.[60]

Given that they are reaching only half the possible market, one could ask why the games-makers don't start to develop material which will interest the girls. Joseph Pereira says that an attempt was made, but that the few games that were aimed at females, "including Echo Dolphin and a series of Barbie games, have been disappointing".[61]

Well, I have heard all this before. It was the catch-cry of so many male publishers in the 1960s and early 1970s when they were asked why they had so few books on women. "We tried one once, and it didn't sell" was the standard response. Or "The wife said she wouldn't buy such a book, so we didn't do it." Well, if she were "the wife" – she would, wouldn't she?

But who was it who made the decision on the particular computer games that were supposed to be for the girls? Were they designed with girls' values in mind, with some knowledge of women's ways of knowing and doing,[62] or did the programmers just ply their skills with Barbie rather than *Mortal Kombat*?

Leila Chang, a recent appointment to HiTech Entertainment, claims that the very little that has been produced for girls has often been a 25-year-old male's idea of what a little girl would like to play. And it has been way off-beam; she is scathing about the behaviour of some of her male colleagues:

> "To a lot of guys in the industry it never occurs to them why their wives never play video games. Women are not interested in activities that reinforce the notion that we should be sitting round waiting for some guy to marry us and let us tend for him in his house and have his babies," Chang said.[63]

---

59. Smith, Internet communication.
60. Joseph Pereira, 1994, "Video Games: The Chips Are Stacked in Favor of Boys", *Wall Street Journal*, reprinted *Age*, 5 April, p. 21.
61. Pereira, "Video Games".
62. See Mary Field Belenky *et al.*, 1986, *Women's Ways of Knowing*, Basic Books, New York.
63. David Cox, 1994, "Women Tearing down the Walls of the Male Games Bastion", *Age*, 6 December, p. 29.

When it is not a case of a woman being rescued by the hero, it is a game in which male fantasies of women abound. The latest from Philips introduces their "sassy" heroine Lieutenant Jessica Darkhill, who pits her skills against outer space warriors, and who conquers and kills no less than the men: but who is meant to be somewhat more stimulating to look at.[64] As Leila Chang says from within the industry: "There are few role models for females in games. Most are created by boys for boys."[65] And some of them don't want any changes.

When Virginia Barrett, along with VNS Matrix, devised an amazing female-oriented computer game and tried to get financial backing for it, reaction from the various companies was undisguisedly negative. Admittedly, this could have had something to do with the particular design of the game.

The women who created the game are not only highly talented artists, they are cheeky and they have a good sense of humour. They have built in some fantastic tricks. Anyone who doesn't know an appropriate feminist response in some circumstances, for example, immediately finds they have been transported right back to the beginning.

One could assume that in dealing with the incredible DNA Sluts (the characters in this narrative), the average male games-maker might have had some difficulty getting beyond first base; he could have well decided that he didn't want to play. And that no one else would either.

All New Gen (the name of the game) is "an artwork of great wit" according to David O'Halloran. "The work pokes fun at the male of the species. Gen is the super heroine of this game and her mission is to destroy Big Daddy Mainframe."[66]

How deliciously tempting. I haven't done any research on the subject, but I suspect that most young women would thoroughly enjoy this game. I know my niece would be delighted with the twist, and would undoubtedly spend a long time learning all the skills that boys are supposed to grasp from the hours spent "manipulating" the controls and the screen. But she probably won't get the opportunity to find this out. Because here we are well and truly into the electronic revolution where games are the pace-setter, and one of the biggest items. Not only have girls been left out: there is little likelihood that they will ever get in on the act.

> It's going to take some initiative by manufacturers, agrees Laura Thornton, a spokeswoman for Capcom USA, a video game maker in Santa Clara, California. "Right now it's a question of the chicken or the egg. Manufacturers say they're not making more games for girls because girls aren't interested, and girls say they aren't interested because manufacturers aren't making more games for girls."[67]

Creating a few games for girls will not be enough, of course. It's quite possible that there will be some young women who don't like *All New Gen*; just as some didn't like

---

64. Trudi McIntosh, 1995, "Cyberspace Heroine Shoots for the Stars", *Australian,* 28 March, p. 23.
65. Quoted in Cox, "Women Tearing Down the Walls of the Male Games Bastion".
66. David O'Halloran, 1993, *Adelaide Advertiser,* 10 November.
67. Pereira, "Video Games".

Enid Blyton or *Anne of Green Gables* in my day. This doesn't mean girls don't like computer games: it means that there isn't a great enough range to capture their diverse interests.

There needs to be a variety of games, CD-ROMs and software available, which excite, thrill, charm and delight girls, and which develop their interest in the personal, in relationships, connections, the way people work. This is what books have been for girls in the past.

Games designer Leila Chang understands this. Even though (in my opinion) her new Barbie games aren't in the same league as *Little Women* and *Seven Little Australians*, they are still concerned with relationships:

> Chang's recently released point and click CD-ROM title *Barbie and her Magical House* involved extensive focus testing with the target group of four- to seven-year-old girls, who requested that Barbie be shown in a variety of settings. (This was very different from the earlier male-designed products such as *Barbie Supermodel*.) In this title (*Supermodel*) Barbie tours the world, in search of fame and fortune. Most people who buy Barbie games tend not to think about such things, according to Chang. She based the CD-ROM title on social activities like horse riding, going to the beach and camping.[68]

But Leila Chang's product is unusual. Just about everywhere else, it's toys for the boys at the moment. They are the ones with the leisure time, the ones who are seen as the market, the ones whom the games-producers are doing their utmost to please. The only problem for women is that it is not a game that we are playing; in more ways than one it is deadly serious. Women, and women's experience, are being excluded:

> About a third of American families have at least one computer, but most of these are purchased and used by males. It may be new technology, but the old rules still apply. In part it's that male–machine bonding thing, reincarnated in the digital age. "Men tend to be seduced by the technologies" says Oliver Stimpel, executive director of the Computer Museum in Boston. "They tend to get into the faster-racing car syndrome," bragging about the size of their discs or the speed of their microprocessors . . . This is not something to be trifled with by mere . . . females, who seem to think that machines were meant to be *used*, like the microwave oven or the dishwasher.[69]

Talking Up the Telephone

One theme that runs through the commentary on the gender gap and computers is that women and men see the computer differently. To put it simply, the consensus is that women use them, whereas men fall in love with them. It is not far removed from the gender-different response to cars.

If computers were seen to be primarily about computing, about solving problems and paying bills, etc., it's clear that women could be persuaded to use them, particularly

---

68. Cox, "Women Tearing down the Walls of the Male Games Bastion".
69. Kantrowitz, "Men, Women and Computers".

if they were presented like other household appliances as saving time. According to Margie Wylie:

> Women usually want to accomplish specific tasks when they sit down at a keyboard. They are looking to save time, which makes them less tolerant of working round snafus in software and hardware, such as messages with endless variations of modem strings, or system configurations. Plus women just don't have the time to waste. According to the Department of Labor, 58% of all American women worked outside the home in 1993, while they also still perform the lion's share of household chores and child care.[70]

That women use computers, while men are ensnared by them is a point that is also made by William Bulkeley in his *Wall Street Journal* article entitled "A Tool for Women, a Plaything for Men". But while he says that part of the pattern is for men to spend more time embracing their computers, he does not mention the issue referred to by so many feminist commentators – namely that women want the thing to work because they are so pushed for time. For men in contrast, computers have already become a leisure pursuit.

> While women use computers, men love them . . . Though women may not like their computers as much as men do, women may actually be better at getting work done on them . . . Men also spend more of their spare time on their computers. Ziff Davis, the computer magazine giant, says the readership of its biggest magazines such as *PC Magazine*, *Computer Shopper* and *PC Week* is 80% male and 20% female. User groups are two thirds or more male. And CompuServe says just 20% of its on-line subscribers are women . . .[71]

Jen St Clair – one of the rare researchers in this area – has drawn attention to the fact that it is the leisure needs of males that play a key role in determining what is on offer in the expansion of the information infrastructure. Assessing the implications of pay television in Australia, her analysis suggests that the "time-rich and cash-poor" members of the community would be the majority of viewers. They would be mostly men, and not from the managerial or professional classes. Her data led her to conclude that "low-cost programming will predominate on pay television networks" with a focus on sport and action films.[72] Hardly a boon to women.

So men's interests are once more met at the expense of women's. This has occurred in a new form where – theoretically at least – there was the chance to cater for both sexes equally. There probably won't be equality of provision on TV, cable, CD-ROM, etc., until there is equality of viewing. And as yet I have not come across the suggestion that, in the interest of gender equal programming, more leisure time should be given to women.

---

70. Wylie, "No Place for Women", p. 4.
71. William M. Bulkeley, 1994,"A Tool for Women, a Plaything for Men", *Wall Street Journal*, reprinted *Age*, 5 April.
72. Errol Simper, 1995, "A Matter of Time as Pay TV Faces Inbuilt Problems of the Market", *Australian*, 18 January, p. 2.

When women are asked about their leisure interests, there are many who say that the little time they get is used mainly to chat with family or friends. Sometimes this takes the form of having a cup of coffee; meeting at home; having a bit of a gossip while watching the kids. It isn't always necessary to meet in person, however.

Talking on the telephone is one of the most commonly reported "non-work" activities of women.

This is of course very interesting. The telephone is an integral part of the information infrastructure, yet women's easy and ready use of this form of technology usually goes unnoted. This is one aspect of information technology where all could be reversed, and it could be said that men *use* the telephone, but women *love* it!

But women's proficiency and pleasure in this machine is never cited as a reason for women being superb with computer networks. That women already spend considerable time in cyberspace (having conversations on the telephone) is almost never taken into account.

Women's love affair with the telephone was rather late developing. As Lana Rakow indicates in *Gender on the Line*, the telephone was not seen as appropriate for women when it first made its appearance. Its initial commercial use was as a primary means of business communication and it was treated very seriously. "Private calls" – that is, those made by women – were charged at a higher rate to discourage frivolous use.

Since then, women have made the telephone very much their own technology. This doesn't mean that they are overwhelmed with a desire to find out how it works, or to tinker with it in their "spare time". They still just want to use it rather than "play" with it. This is one reason that Lana Rakow thinks it is misleading to see women's time on the telephone as simply a leisure activity. She insists that much of what women are doing is still work: "the telephone provides a network for gender work (social practices that create and sustain individuals as women or men) and gendered work (productive activity assigned to women)."[73]

Women do relationship work on the telephone. It is their means of keeping in touch, of maintaining contacts, of building and supporting a network – a cyber-community, no less.

This is the work of human maintenance, of emotional management, of breaking down isolation, and of keeping communication flowing. And while it is often social work, there is no denying that it can also be fruitful and enjoyable. In contrast to the bad press that "gossiping" often gets, it can be a most positive activity; it can make those who gossip feel immeasurably better. (Note that the word *gossip* once meant a comforter who talked you through the birth, who talked away pain and made you feel relaxed. This is still one of its functions, although it is no longer confined to times of childbirth.)[74]

In her study *Women and the Telephone in Australia* (1989), Ann Moyal documents the "pots and pans" emotional links, and the vital social support system that women's

---

73. Lana Rakow, 1992, *Gender on the Line*, University of Illinois Press, Urbana and Chicago, p. 1.
74. Cheris Kramarae *et al.* (eds), 1995, *A Feminist Dictionary*, Pandora, London.

time on the telephone sustains. And she points to the "economic importance of this support system" which is now being recognised.[75] Not that the telephone was seen as such a useful way to maintain relationships when it was first invented. The general objections which were raised then have a familiar ring now as we use the telephone lines (and modems) to connect one computer to the next.

The introduction of the telephone was accompanied by prophecies of doom and gloom about the end of human communication when there was no longer the need to talk face to face. It would be the end of nuance, of irony, of feedback, if you couldn't physically see the person to whom you were talking. All manner of mistakes would be made because you might not know whether the other person was male or female, or of what class or race.

There was even apprehension about the way people would behave when they could communicate anonymously. To some extent, this was justified. Until the advent of answering machines, the obscene telephone call was quite a problem; it introduced yet another means which allowed sexual harassment.[76]

But far from reducing communication opportunities, the telephone provides an excellent example of the way in which technology can extend the range and the repertoire. The telephone has added another possibility and has helped forge new networks and communities.

And this is how I would urge women to start seeing the computer: as a means of communicating, of plugging into the biggest network to be devised, of making and maintaining friendships and contacts.

Computers are for nattering on the net.

A computer connected to the information superhighway is more like a telephone (with the added benefit of pictures and text) than it is like an adding machine. Like the telephone, the connected computer is just crying out for women to use it.

The computer viewed in this way will soon – like the telephone – cease to look like technology. There would be few women who think of their telephone as a machine – the technology has become invisible, as will that of the computer.

Women will be drawn in through an emphasis on the communication potential of the computer. Once women can see that it is dead easy to natter on the net – to reach people all around the world, to consult bulletin boards, to "meet" in cafes and houses and art galleries without leaving home – there will be no stopping them.

The only obstacle that they will have to contend with is the men who are already there; the men who have written the rules of the road. There will be no shortage of intimidatory and dangerous drivers at large, trying to scare women away. But that's another story.

---

75. Ann Moyal, 1995, "Crisis in Communication Research", paper presented at the Communication Research Centre, The Informationless Society Seminar, 16 February, p. 6.

76. While I have done no research in the area, I suspect that the invention of the postal system met with many such protests – that it would be the end of human communication, and that people would behave badly when they could send messages anonymously. It was poison pen letters that were the problem in that medium.

Male Menace on the Superhighway

Writing the Road Rules

In the real world, men dominate communication.[77]

Men talk more often, they talk for longer periods, they adopt "centring" positions (forcing females to hover around); men define the topic, assume the legitimacy of their own view, and override women who do not see the world in their terms.

Much of this dominant status is achieved by interrupting and correcting. Men perform up to 99 per cent of the interruptions in mixed-sex conversations, and one of their most common ploys for taking the floor is to interrupt women with a version of the comment "What you mean is . . .". That is, men take over the topic by correcting the woman; they redefine her meaning in their own terms.

(Of course, for the woman, the last thing she may "mean" is the male takeover of her words; but unless she wants to be seen as rude and aggressive, she will probably, politely, acquiesce.)

Not all men do this. Not all men choose to stick by the rules. (Not all women go along with them either.) But the fact that some men do not conform to the conversational code does not mean that the code ceases to operate. Indeed, it can be quite the reverse. In choosing not to play by the rules, not to interrupt, dominate, or talk over a woman, a man can be so noticeable that his behaviour serves as a reminder of the ground rules that he is breaking, and this entrenches them still further.

For the last two decades, at least, there has been overwhelming evidence that men are the talkers in real life. While women can be very aware of this, and quite stroppy and assertive in their own right, their chances of changing this communication pattern are not high. One reason for this lies in the way the rules are policed and reinforced.

A woman who breaks the rules and tries to get a fair share of the time and the topic is invariably penalised for her "infringement". Unlike a man who can find himself praised for his attempt to promote more equal conversation habits, the woman who shares the same aim is likely to find herself ridiculed, abused, punished. Terms such as *domineering*, *bitch*, *bossy*, and *castrating* have all been used to describe women who take more than 30 per cent of the conversation space.

Such tactics play a crucial role in keeping women in their linguistic place.

As the values of the real world can simply be transferred to the virtual world, we shouldn't be surprised to find much the same pattern in cyberspace. The only difference between the real world and the virtual world is that, if anything, the male domination of cyberspace *is worse!*

In a rare – but nonetheless highly illuminating – study on women and men on the net, Susan Herring, Deborah Johnson and Tamra DiBenedetto have managed to document some of the rules of mixed-sex conversation in virtual reality. The figures that they have on male/female contributions show just how marginalised women have become.

---

77. For further discussion, see Dale Spender, 1980, *Man Made Language*, Pandora, London.

Even women linguists have trouble finding a space. On the professional electronic discussion list LINGUIST, women contributed 20 per cent of the postings; they were least involved when males "intellectualised" a discussion ("What you mean is . . ."). It is interesting to note that, when surveyed, both sexes "reported being intimidated by the bombastic and adversarial postings of a small minority of male contributors who effectively dominated the discussions".[78]

So what's new in cyberspace?

In trying to work out what was happening on the net, Susan Herring *et al.* decided to keep track of the conversation on Megabyte University (MBU) for a period of five weeks. This was a small friendly list, where feminism enjoyed some influence. The researchers believed that women would get a fairer share of the postings in this non-adversarial environment.

The results were disappointing. Men retained 70 per cent of the conversational space.

In the course of their research moreover, Susan Herring *et al.* monitored an incident which revealed that there is one rule for women and quite another for men in the information superhighway code as it stands at the moment.

At one stage in the five-week period, a feminist topic was raised, and for two consecutive days, women posted more messages than men. Despite the fact that, on the 33 other days for which records were kept, it was men who took up most of the space, there was an angry response on the part of some men when women took a two-day turn.

Accusations came from the men that they were being silenced. Some even threatened to "unsubscribe" from the list. One man wrote the lengthiest message of all – 1098 words – protesting that the women were "shutting up" the men with their vituperation and insults. This surprised the researchers:

> In looking back over the messages posted during the immediately preceding days, however, we find little evidence of a vituperative tone. With one exception, the contributions of the women appear to be aimed at furthering communication; they raise questions about the interaction at hand . . . explain their own views and encourage others to respond in kind. The only message indisputably negative in tone was posted by the man (who started the issue and who wrote the 1098-word protest) . . . In it, he accuses women on the list of "posting without thinking their contribution through carefully first", of levelling charges (rather than questions) at the men and in general of "bashing", "guilt tripping" and "bullying" men who don't toe a strict feminist line. A man who overtly sided with the women also comes under attack; he is accused of betraying his brothers out of feminist-induced guilt.[79]

There are quite a few issues here. The first is that when men do it, when men take up most of the space – it's OK. There's no problem. But when women do it, there are men who become emotionally abusive and defensive.[80]

---

78. Herring, Johnson, & DiBenedetto, "Participation in Electronic Discourse in a Feminist Field". For further discussion, see p. 225.
79. Herring, Johnson, & DiBenedetto, "Participation in Electronic Discourse in a Feminist Field".
80. There is further discussion about this on pp. 224–8.

Some men get angry when they aren't in the driver's seat. They make threats. They engage in intimidatory behaviour in the attempt to wrest back the controls.

That this tactic should work in cyberspace is to some extent unanticipated. It might have been expected that, because there is no threat of direct physical abuse or violence in virtual conversations, women would be able to claim equal rights. But this is not the case. For it seems that as the physical risk has been reduced, the element of psychological threat and intimidation has risen accordingly, to the point where it can be said that, under the present code for the information superhighway, men control the rights to participation.

This doesn't mean that women can't set out to take up their share of the road. It doesn't mean that they can't insist on a new code that respects their rights. The potential for equality is there in the virtual world as it is in real life. But with so few women in a position to be heard, and heeded, the prospects for improvement are grim. As Maureen Ebben and Cheris Kramarae have commented: "we acknowledge the powerful potential of new information technologies, but we are also painfully aware that many of these potentials have yet to be fully realized for women".[81]

While more research is needed, there can be no mistake about the findings that have been presented so far. Women have fewer rights to discourse in cyberspace, and the penalties for challenging male control could be even harsher than in real life. There are some obvious explanations for this state of affairs.

First of all, the ratio of men to women is so much greater in the virtual world. (As the study of James Pitkow and Mimi Recker indicates, the proportion of women on-line could be not much more than 5 per cent.)[82] Another factor is the anonymity; there is ample evidence which suggests that users can behave in a much more abusive and crudely coercive way when they are anonymous, than they would in a face-to-face situation.

In her study of women's response to communication on the net, Randye Jones found that there was considerable awareness of the opportunities for menace and threat. Women commented at length about the tactic of "flaming" – of intimidating someone whose contribution you don't like:

> When you are speaking face to face with someone, no matter how angry you may get at what the person says, their *physical* presence may act as a deterrent to "flaming" them orally:
>   a) they might be big enough to belt you one;
>   b) you might not *really* want to hurt their feelings;
> – thus, there is an element of self control that gets into most face to face conversations. The fact that fights often break out (among males) in *bars* – when the inhibitions to violent language and nastiness have been dampened (if not destroyed) by alcohol – is evidence for this view.[83]

---

81. Ebben & Kramarae, "Women and Information Technologies: Creating a Cyberspace of Our Own", p. 16.
82. Carli, "A Designer's Guide to the Internet", p. 27.

"Flaming" is not the only indication that people can behave more rudely on the net than in face-to-face communication. For example, males who in real life probably wouldn't dream of butting in on a group of women, appear to have no compunction about posting messages to women-only groups on the net. A significant number of these postings are sexually explicit and abusive.[84]

It is because men have dominated communication that even in real life women have tried to establish forums where women's rules prevail: where women are in charge of the conversation and free to exchange ideas. The women's caucuses of professional societies, women's sections of trade unions, women's student organisations, etc., are all instances of women trying to create a space where they can present and analyse their own experience. Similar groups have now been set up on the information super-highway; and with predictable consequences.

> Even in feminist forums where women are ostensibly more interested and expert, men dominate the conversation. In a 1993 study of the newsgroup alt.feminism, Laurel Sutton found that men contributed 74% of the postings. (Women accounted for 17% and the remaining 9% were contributed by people of unknown gender.)[85]

This is just one of the significant number of studies which shows that women get even less conversational space in the virtual world than they do in real life. This is extremely worrying. At this very moment, as the road rules for the superhighway are being written and worked out, there is no critical mass of women involved to ensure that the highway code reflects some of their priorities and interests.

"Netiquette" – as the new code is called – is another good example of the way men get there first and then stand guard at the gateway; their rules of entry are that you have to play their way if you want to be allowed on the road. It could even be that there is a willingness to open the gate to as many women as men, as long as they play according to the road rules that the men have already set up.

Susan Herring has undertaken an analysis of some of the rules of the new netiquette, and she has found this definite male bias. She has an apt way of describing the values that apply:

> To simplify, net rules tend to look like the opening scenes of an old-fashioned boxing sequence on a late night movie: "No hitting below the belt, no fighting dirty; may the best man win. Ding. Ding." Most netiquette statements acknowledge that the going can get rough, but the answer is invariably that if you can't take the heat, stay out of the kitchen.[86]

There are many women who don't want to play this way; who don't want to conduct conversation according to these rules. As a result, even when they are on-line, they

---

83. Randye Jones, 1993, "Netters Behavior and Flame Wars", on-line communication, September.
84. There is much more discussion about this sort of behaviour below.
85. Wylie, "No Place for Women", p. 4.
86. Quoted in Wylie, "No Place for Women", p. 6.

may make no contribution to the male-dominated communication. "Men's behavior puts women off," claims Margie Wylie. "When men attack women's posts (by flaming, intellectualising, or posting lengthy line-by-line rebuttals of women's messages) women retreat into silence."[87]

Such silence is more than understandable. Women are being kept out of cyber-communication with an electronic version of interruption and intimidation. If you want additional examples of the more aggressive practices of cyberspace, there is no need to go further than the habit of line-by-line rebuttal of women's arguments; this is but another version of the "correction" statement, "What you mean is . . .", taken to the extreme.

Women are being silenced on the net in a number of ways. There is the silence of women who withdraw because they are shocked, fed up, threatened or distressed. Then, too, there is the silence that is imposed – the silence that goes with women not being able to get a word in, prevented from raising the topics that are of concern to them. But there's also the silence that confronts women if and when they do venture to comment on-line. In society in general, and in cyberspace in particular, it is abundantly clear that women's words count for much less than men's:

> More often than not . . . women's ideas are simply met by silence from men and women alike. A female-initiated subject gets roughly less than a third of the replies that a male-initiated thread does . . . And even women-initiated threads that survive are taken over by men . . . so the eventual fate of the conversation isn't guided by women. What's more, men tend to post longer messages more frequently.[88]

There is not a lot that women can do about this gross inequality. Protests are likely to lead to further flaming and intimidation; you know the score – "if you can't take the heat, stay out of the kitchen". Yet any attempt to create women-only space is just as fraught; if it is not over-run by males, it is likely to be labelled as sexist and to attract further flaming and abuse. "When women retreat to all-female forums," says Margie Wylie, "many men accuse women of impinging on their freedom of speech by trying to exclude them from the conversation."[89]

It is interesting that the principle of free speech should be so vigorously championed by so many men. There is considerable evidence that there are those among them who think that free speech applies only to members of their own sex. That is, they are free to say anything they like – but it is not on to say critical things about them. Such are the benefits for those who have the power to write the rules in their own interest. (For further discussion, see pp. 224–8.)

This double standard has not been lost on the women who are critiquing the behaviour of men on the net.

---

87. Wylie, "No Place for Women", p. 4.

88. Wylie, "No Place for Women", p. 4.

89. Wylie, "No Place for Women", p. 4.

> While the Internet is ostensibly free of any censorship, women find that by the very nature of their language, what they have to say is not only unheard, but does not fall within the parameters of acceptable discourse.[90]

This is what men can achieve while they are in charge of the rules. They can develop a code that gives them right of way; and with so few women in cyberspace there is little serious protest to establish that this is not fair. Women aren't allowed to do what men do – and they aren't allowed to develop a code of their own.

The studies that have been done on communication on the net make it clear that it's more a male monologue than a mixed-sex conversation.[91] The discourse is male; the style is adversarial. The premises are winning or losing. Despite the enormous potential of the net to be a network – to promote egalitarian, cooperative communication exchanges – the virtual reality is one where aggression, intimidation and plain macho-mode prevail. As some perceptive researchers have noted, the behaviour is not so different from that of aggressive men behind the wheel of the car on the road.

Mike Holderness is a commentator on the new technologies, and one male who is acutely aware of the pitfalls of the system. He has even tried to prepare beginners (the "newbies") so that they can deal with the shock that goes with their entry to on-line communication; he provides guidelines intended to ward off the bullying and the blows:

> You get connected to the Internet. You are excited about being plugged into probably the greatest information resource ever created. Twenty or 30 million people are just out there waiting to communicate with you! You want to know everything there is to know about . . . Baroque music.
>
> You pick two dozen Usenet news groups out of the 3000 or more available to you and send a message to each: "Can anyone tell me where I get information about Baroque music?" Next time you connect your mailbox is full of messages from the four corners of the earth. Joy. Communication. But message after message is abusive. And you are being admonished about things you are not even sure you can pronounce. Spamming? FAQ? RTFM?
>
> Welcome to Usenet. You have been flamed.[92]

Mike Holderness also notes that newbies don't stay newbies for very long. The newcomers quickly learn the highway code, and then are keen to strut their stuff with the next lot of newcomers. One of the worst things about the net, he comments acidly, is encountering "the boys who have been using it for three months and feel the need to correct the newbies".[93]

---

90. Wylie, "No Place for Women", p. 4.
91. See Cheris Kramarae & H. Jeanie Taylor. "Women and Men on Electronic Networks: A Conversation or a Monologue?", in Taylor, Kramarae & Ebben (eds), *Women, Information Technology, and Scholarship*, pp. 52–61.
92. Mike Holderness, 1994, "Netiquette for the Novices", *Guardian*, reprinted *Age, Green Guide*, 13 October, p. 25. For anyone who is interested: FAQ is "frequently asked question" while RTFM stands for "Read The F—— Manual".
93. Holderness, "Netiquette for the Novices".

Like Mike Holderness, Lynda Davies is also acutely critical of the macho nature of virtual communication. She doesn't only witness the on-line tactics of dominance – she sees them face to face in the computer labs where she does her research. She describes an average day in an average computer classroom where men dominate:

> On the computer screens, some of the students are writing on news groups, which are communication channels for computer enthusiasts to talk to each other through electronic forms of writing. The writing tends to take the form of adoption of character identities, pseudonyms, and writers communicate with similar other groups around the world. Once in a while someone will shout out telling others to net into a group news item which is dealing with some form of discussion, often about continuing contributions to a joke category. The joke categories are most often explicitly sexual, or about some form of killing or mutilation. War and sexuality are forms most frequently net talked. . .The sexuality is almost always phallocentric, giving the image that the aggressive penis centred activities are the key attributes of the hero figures.[94]

Lynda Davies makes no bones about the fact that she is outraged by the behaviour of the computer-boys in real life, and by what they do on-line. She sees this macho behaviour as an explicit attempt to scare women off so that the boys can keep the territory to themselves. There can be no doubt that this strategy is working:

> despite the early explorations and sometimes ground breaking work women have done in computing, women in many rapidly technologizing fields have felt increasingly discouraged by the discursive practices they encounter on the "nets" . . . These studies serve to confirm anecdotal evidence from professional women who have participated actively in network culture: women frequently feel ignored, silenced, even abused in electronic conversations.[95]

What is most deplorable, as far as Lynda Davies is concerned, is that it is in universities that much of this grossly sexist and intimidatory behaviour is flourishing and is being condoned. So macho is the entire virtual world that she knows, that she has found it almost impossible to draw the line between legitimate computer use and totally unacceptable abusive behaviour.

As she sees it, the technical people of the future are being trained within the computer science labs. At the moment they are being trained in appalling misogynist practices that would not be tolerated anywhere else on the real-life campus. She describes some of the everyday realities of the computer science classroom where the makers of today's and tomorrow's codes are being trained:

> This is the regimental training ground for computer technologists. . . . This is the same training ground in which debates on pornography, child sexual abuse, sexual deviance in all forms are frequently conducted. Knowing these jokes (and accepting them) is a method of entry into he social world of the regiments of technologists. Trophies are taken in the form

---

94. Lynda Davies, "The Gendered Language of Technology".
95. Kaplan & Farrell, "Weavers of Webs".

of "gifts" (pictures) which are "downloaded" (dumped onto portable disks from the computer networks) and portrayed on home computers, usually within some packaging making it more difficult for parents to accidentally come across these "gifts". The gifts are usually nude shots of females or of explicit sexual poses; they are the *Playboy* centrefold of the technology regiments.[96]

It is extraordinarily difficult for young women to know how to behave in the computer classrooms and when they are on-line. The dominance of men and of extreme macho-values ensures that any woman who does stay has to find her own means of accommodation with the brutalising nature of the medium as it has been structured.

Some women try to be invisible. Some try to go along with the ribald and sexist "humour". Most just want to pull out. But this isn't always possible. As the world moves more towards on-line communication, women have fewer and fewer choices about whether or not they will participate. For many professional women, staying out of cyberspace is not an option. But becoming a speaker can exact a terrible price:

> A highly respected lawyer and electronic copyright expert who hosted an on-line discussion on a commercial service in late 1994 reported that the first question her all-male audience submitted was, "When is your menstrual period?" After the moderator threw the question out as inappropriate, other participants inquired about her marital status, her looks, her hometown. Not a single question was raised within her area of expertise.[97]

Why would she want to repeat such an experience? Where can she go to protest at the rules or to have them changed? What chance is there of men "seeing the error of their ways" within the present system? And how can all these computer whizzes uphold the principle of free speech in cyberspace, when for women there is not the remotest possibility of a fair hearing?

Even the language of the road rules puts women off: "the use of such terms as 'abort', 'chaining', 'thrashing', 'execute', 'head crash', and 'kill' provides an 'undercurrent of violence' producing an imagery of sex and violence which is contrary to girls' social conditioning".[98]

More worrying, perhaps, is the way that women once again are being made responsible for an area – and an attitude – over which they have no control. One of the arguments for co-ed schools was that the girls would "civilise" the boys. (It is interesting to note that no such role is claimed for those oppressed by class or race. I have never heard the rationalisation that the presence of blacks or the working class would civilise the sensibilities of the white middle class. Quite the reverse.) But the same argument for women improving the behaviour of men is surfacing again; only this time it is that we need women to "clean up" the sexism and sleaze of cyberspace.

---

96. Davies, "The Gendered Language of Technology".
97. Wylie, "No Place for Women", p. 5.
98. Cole, Conlon, Jackson & Welch, "Information Technology and Gender: Problems and Proposals", p. 81.

The existence of so much porn and sexually explicit material on the net has even spawned [*sic*] a theory among academics that unless more women are encouraged to enter cyberspace, the morals and behavioural standards by which it operates will be set exclusively; by men.[99]

Because there is so much aggro to contend with, among the advice that is currently handed out to those who are new on the net, is the suggestion that they should stay in the shadows and become familiar with the rules before taking to the road. The term that is used to describe this practice is *lurking*. It's a term that is unlikely to have positive connotations for women; the very idea that cyberspace is inhabited with "lurkers" would be enough to make many women apprehensive and to send some hurrying for cover.

The fact of the matter, however, is that women cannot afford to stay away. More and more of the world's business – along with education, entertainment and consumerism – will be conducted on-line. And women will have to become part of cyber-society. This is why women have to insist on the right to make up at least half the rules in their own interest. It is why they have to demand the right to a more hospitable and respectful environment.

Women cannot expect that their call will be understood or their claim treated sympathetically. The following is a letter to the editor of the computer section in *The Australian*, a response to women's demands for a better deal on the net.

Nothing does more to define feminists as a pathetic, ineffectual collection of moaners than your article, "Women take the feminist flame to cyberspace" (14.2.95).

For two decades guys have been spending their nights logging onto bulletin boards and other computer networks.

For this they were dismissed as pimply-faced adolescent nerds – but it turns out they were defining and pioneering the use of one of the most crucial and important technologies of our day.

This has finally occurred to the idiot feminists and the male-to-female imbalance on the Net has become a "gender equity problem" – another opportunity for women to fight for equality in our patriarchal, male-dominated, sexist society.

Get serious. How stupid do we look?

"Grrrls need modems" is the catch cry.

What girls might find more useful is the smallest amount of technological and scientific vision, or the guts and strength of character to persevere with an interest in the face of a "consensus" that says you're wasting your time.

Michael Lines, Keperra, Qld.[100]

Despite such attitudes, the netiquette of cyberspace is not as yet entrenched. The scope to make changes still exists; but the opportunity will soon pass. Without change, the exclusion of women could be institutionalised.

---

99. Craig Johnstone, 1995, "Surfin' for Sex", *Courier Mail*, 5 April, p. 9.

100. *The Australian*, 21 March 1995, p. 24.

Sexual Harassment

Cyberspace is a public place and women are at risk in it, as they are in every other public place. But whereas girls and young women have been socialised to be street-wise in the real world – to detect danger signs and to be able to protect and defend themselves – such training has not been readily available for cyber-society. It's new territory; the women who are there are still finding their way. As yet they are not in a position to pass on the benefit of their experience.

This is most unfortunate. Some researchers suggest that the need for self-defence strategies is even greater in cyberspace: that as sexual harassment in the public arena and at the workplace is being checked, it is running rampant on-line.

From schoolgirls to seasoned professional women, the virtual reality is the same. There is no female who has worked on a networked system who has not been subjected to harassment, flaming, or other intimidatory tactics.

It's bad enough for women who are using a commercial service, paying good money for their connection to the virtual world. But at this stage the majority of users (particularly in Australia) are linked through an educational network. That such a system is used to harass and intimidate faculty and students is a gross misuse of resources and power. In the end, it is also harmful to the institutions.

> Sexual harassment on the networks is a problem being reported by many women at many sites. The forms of sexual harassment vary. In some cases women on the networks experience the types of sexist jokes and limericks that were once part of many classroom lectures but now are increasingly discouraged by university administrators. In other cases, women's complaints about the harassment they experience lead to hostile comments about the women. Women who want to be respected scholars (which means avoiding being known as a trouble-maker, or as someone who questions the actions of respected men in the field) will be very hesitant about publicly pointing to sexist behavior. This silencing has serious consequences for interaction on individual nets, for disciplines, and for the academy in general.[101]

In one college computer-writing classroom that I observed in the United States, sexually explicit material kept appearing on the screens and the women students were taunted with snide and sniggering remarks. "None of these boys would dream of bring-ing *Playboy* into the classroom and putting it on a girl's desk," the lecturer informed me. "But they think it is fine and funny to put centrefolds in the screen and to see the girls squirm."

This instance in itself is an indication of the way things are getting worse for girls. Not so long ago, writing classrooms were where girls excelled: composition was their forte. The ethos was one where they were comfortable; the space was relatively safe. But since then, the classroom has been transformed into "computers and writing"; and it has become male territory in the process. Composition classrooms can now be hostile environments for girls, and sexual harassment can be the means of communication.

---

101. Cheris Kramarae & H. Jeanie Taylor, 1993, "Women and Men on Electronic Networks: A Conver-sation or a Monologue?", in Taylor, Kramarae & Ebben (eds), *Women, Information Technology, and Scholarship*, p. 56.

Every co-ed college, and every on-line school that I know of, has had to deal with an extraordinary level of sexual harassment. At one school, fourteen-year-old boys were posting messages all over the world declaring that particular females in their class were whores, slags, and much more, and that they would do anything for anybody. When some of these fourteen-year-old girls then logged on, they found the most vile e-mail messages it is possible to imagine. Naturally, they were upset by this confrontation. They were not only psychologically disturbed by the violations and the threats, but a few of them became phobic about using the computer at all. In much the same way, those of us who were the victims of obscene telephone calls in the past came to dread the telephone.

Sexual harassment has often been referred to as the systematic means of keeping women out of male territory, and this is certainly how it works in cyberspace. Young women who are distressed by the face-to-face behaviour of males in classrooms and computer labs, and who are shocked to the core by their early experiences on the net, aren't going to be keen to become the computer whizzes of the twenty-first century. This is why sexual harassment should be seen for what it is: the "terrorist" tactic used by some men to drive women away from the centre of wealth and power.

In writing this section on "guerilla warfare", I face something of a double bind. I believe that few people could conceptualise the sort of abusive messages that flow through cyberspace, but it doesn't give me any satisfaction to quote some of the grossly offensive incidents and to give them the force that the writer intended. I baulk at the idea of acknowledging that these threats are effective; such feedback adds to the power of the harasser. But silence is no longer an option.

Rosie Cross, for example, has sent me copies of some of the messages that she has found routinely on her screen. They are harrowing in the extreme. "The only pre-requisite for abuse and harassment on line is to be female," Rosie Cross remarks, and provides the evidence:

> I logged onto a board, using the name Emily, and immediately I was pounced upon by requests to go out and fill in my registration in order that the boys can "fuck me".

While I have nothing but admiration for Rosie Cross's stand in going public, I also know that abuse like this takes its toll:

> You are welcome to use these examples: unfortunately I deleted my Valentine's Day Message this year . . . Some rather nasty testosteroid left me a message saying that I would have my throat cut and be gangbanged, "fucked to death", for a Valentine's Day treat. I found this message far too disturbing even to keep as evidence.[102]

Not many women have gone as public as Rosie Cross – and you can't blame them. The temptation would be to delete such messages from the screen and to try to forget

---

102. Rosie Cross, 1994, private communication.

them as quickly as possible. (I even wanted to omit them from this coverage.) But there is little chance that the whole sorry business will come to an end if the evidence keeps being blotted out.

When I asked women to contact me about the harassment they have been subjected to on-line, more than a few responded with the information that they had difficulty getting their male partners to appreciate the horrific nature of much of the abuse. At least, that was the case until they persuaded the men to log on as women. When the men used female pseudonyms, without exception they were appalled. "I genuinely had no idea it was as bad as this," said one, who couldn't believe the number and the nature of the postings that he got.

One woman (whose partner was on the receiving end of some of the offensive messages) said that she and her female colleagues had been complaining to the authorities for ages that something should be done about the level of harassment but to no avail. Yet when her male partner was subjected to the same harassment and he protested, his grievance was taken very seriously and dealt with very quickly. (She was pleased that something was done about the harassment, but she drew my attention to the fact that the word of a man – who had been treated like a woman – counted for much more than the word of a number of women.)

In the interest of providing a learning experience for male teachers and faculty, logging on to the Internet with a female pseudonym should be made mandatory. At least then they would have some idea of what women are facing. This is the experience of one male university teacher:

> When I first began using an electronic network about 1981, I had a gender-neutral log-on ID. Before I learned how to set "no-break", I was habitually plagued late at night by young testosterone-laden males who broke in wanting to know if I were an "M or F?"
>
> When I flashed "M", the sender departed, only to be replaced by another flasher with the same question. Only once was the sender a female, as she later revealed in person. On those occasions when I was feeling malicious, I would send back an "F". I was amazed at the simplicity and coarseness of the pickup lines. In discussing this with female students, I learned that such interruptions were common for them, and "no-break" was the second on-line command they learned ("log-off intact" was the first).[103]

Ensuring that men know what it is like to be treated like women is one thing; getting them to do something about it, however, can be quite another. It is simply not possible to rely on the university faculty to ensure that harassment is kept out of cyberspace. As Lynda Davies has indicated, some of the systems operators and computer science staff are more than aware of what is going on. They could even be said to be facilitating it in some circumstances.

Although some turn a blind eye, or even contribute to a culture in which such unacceptable behaviour is encouraged, there are others who genuinely want to bring

---

103. Bay Area Women in Telecommunications (BAWIT), "Gender Issues in Online Communications", CUD ftp site in <pub/cud/papers/gender-issues>. *No-break* and *log-off intact* are means of ensuring that no one can break into your text while you are on-line.

this practice to an end. But when it comes to dealing with the problem, they just don't know how to go about it.

Mike Gigante of the Advanced Computer Graphic Centre at RMIT says of the harassment on the net, that in real life no one would get away with it:

> People tend to be far more abusive on bulletin boards or e-mail than they would face to face. Some of these exchanges I have seen on these news groups have been outrageous. If they said it face to face the person saying it would be sued in court with libel and slander suits.[104]

How are women to deal with this? Where to start is even a perplexing issue. Mike Gigante is among those who believe that there has to be a law against it – that sexual harassment has to be made a crime on the net as it is in real life. Yet in order to get to this stage, the wider society has to know, and to condemn, what is going on.

While there are no guaranteed solutions, it seems to me that one of the most constructive strategies is to go public. Just as sexual harassment in the real world had to be named, just as the silence and taboo had to be broken at work and school, so too, in cyberspace there is a pressing need for women to state categorically what is happening and why it must stop. When sexual harassment on the information superhighway moves to the top of the public agenda, maybe then we can expect effective means of dealing with the offenders.

One young women's magazine – *Sassy* – has made a contribution in this direction. Aware of some of the harassment that women were facing, one of the young reporters, "Margie", did her own research; she interviewed other women of her generation who had been abused on-line. Margie then wrote up some of the experiences, and they don't make pleasant reading.

> I asked for specific instances of the difficulty of being an XX chromosome on line. "I was on MUD [Multi-User Dungeons: interactive computer games] a few weeks ago and somebody was like, 'Oh, I heard you were real easy, I heard you spread like peanut butter.' Subtle. And frequently, you'll get paged by somebody who just goes through paging all the girls, seeing who they can get to talk dirty to them . . . there's always that person who pages you 40 times in a row. Which happened to me. I'm like, what's your problem?! I wrote a note to the sysop [systems operator], who e-mailed me back saying it was my own personal problem. I was totally aggravated. I ended up sending mail to this person saying 'Look, I consider this harassment', and I got this letter back from him saying 'You're a bitch'."

The conclusion that Margie came to was that for girls to survive in cyberspace was often just too difficult; it was easier to quit.

> We're women in a society that has given us claim to certain territories and tried very hard to remove us from others. If you're a girl on the net you get sexual advances, you get told you're not as good as the guys. This world is seen as a private men's club (or boys, as the

---

104. Joanne Painter, 1993, "Uni to Look into Sexual Harassment via Computer", *Age*, 5 November, p. 8.

case may be). Maybe we deal with enough shit on the street that the last thing we want to do when we get home is turn on the computer so some 15-year-old boy can proposition us for MUD sex . . .[105]

It might be difficult to police sexual harassment on commercial servers, but there is no doubt that this sort of harassment should not be allowed in educational institutions or the networks of public libraries. It is not just a matter of the misuse of resources; educational institutions have a duty of care to young people, and therefore cannot allow them to be exposed to such abuse. But as the research of Lynda Davies suggests – there is little likelihood that some of the administrators and staff in computer science labs will put their own house in order. Not that the problem of using educational networks for harassment is confined to computer science personnel (or that all men in computer science are beyond the pale, for that matter).

Sexual harassment on campus has been a major issue for the last decade. (There have been plays and movies about it.) The possibility that male university teachers will use their position to put sexual pressure on female students has been widely acknowledged, and in most educational institutions there are grievance procedures and sexual harassment counsellors as part of the policy designed to put an end to this real-life practice. But like all other attributes of real life, this heinous habit can be transferred to cyber-society. As Mike Holderness reports, the sexually exploitative relationship between teachers and students is now raising its ugly head in virtual reality:

In a university computer room near London, Petra struggled to finish an overdue essay. A message appeared on her screen: a member of staff wanted a "conf" – an on-line discussion.

"It started fairly harmlessly but soon moved into questions about what I was wearing, what I looked like, my sexual preferences," she said.

The staff member implied he could use his computer access privileges to track down Petra (not her real name) to "show me why feminists were wrong about sex, that women really like to be taken over," she said.

"This abuse of authority was made still worse by the fact that he knew how to send me messages whenever I was logged on and I was too computer inexperienced to know how to stop them. I couldn't use the computer for months and consequently got behind in all my classes that involved their use."[106]

There is no doubt that we are moving closer towards a cyberspace education system where teachers will be able to contact students on-line: teaching, advising, marking and assessing will all be handled at the computer workstation. But in order to be a student and to participate in this educational process, you will have to be "wired up". You will have to be computer-competent to get on to the information superhighway.

What is going to happen if women students just don't want the hassle of taking to the road? What if they decide that it's just too dangerous, and they won't log on? What sort

---

105. Margie, 1994, "Hi Girlz, See You in Cyberspace", *Sassy*, May, p. 73.
106. Mike Holderness, 1994, "On-line Harassment Slips through the Net", *The Australian*, 19 January, p. 18.

of an education system will this be? Will it be like it was after the print revolution: with an expanding and exciting education system for boys, while the girls get left behind?

If – because of threat – women refuse to participate, they will be cutting themselves off from the information medium. The parallel would be the people who, in a print-based society, could not read, and who were not full members of the community; people who could not keep up in terms of education, occupation and self-realisation. Second-class citizens, no less.

Violence against women has been perceived as a form of sexual terrorism – a threat which keeps all women in their place. There is no doubt that the witch hunts of the early print period were a form of sexual terrorism which suppressed women's knowledge and made them frightened to demonstrate any intellectual competence. But if it took the persecution of women as witches to keep women out of education and the information medium five hundred years ago, much the same goal is being achieved today through the terrorism of systematic sexual harassment.

There is even rape.

In a recent article in the *Sydney Morning Herald Good Weekend Magazine*, the reality presented went one step further – from sexual harassment to data rape.

A woman in Pennsylvania who called herself Starsinger was logged on to a MOO when on her screen (and on the screen of everyone else who was playing at the time) came the words: "As if against her will, Starsinger jabs a steak knife up her ass, causing immense joy. You hear Mr Bungle laughing evilly in the distance."

This was data rape by Mr Bungle. There's much more. I won't go on. You get the picture.

This data rape led to serious protest and ongoing controversy. Some of the women involved wanted the culprit to be put off the net (toaded). But this went against the principles of those who were determined to defend free speech in cyberspace. (There are numerous groups and societies devoted to ensuring the continued "anarchy" of the net; they are adamantly opposed to any form of regulation.) A typical response from the free-speechers was "If you don't like what you see on your screen, then quit the program."

Such arguments have been used extensively with television of course: if you don't want to watch what's on, then *turn it off!*

But the big difference between flipping the switch on the TV and quitting from the computer screen is that it is not *you* who are at risk when watching the television. The television viewer is watching a scenario in which others are being mistreated. It is not the viewer's character, identity or vulnerability that is being publicly displayed and publicly violated on the television program.

In contrast, it very definitely is your "self" on the computer screen. The MOO or MUD player has created a self, and there is much personal identity vested in the virtual portrayal. What's more, turning off the computer won't make any difference to your fate. It will just mean that you don't know what is happening to this part of your self, although other people who are plugged into the game's "living room" will be privy to every detail of your assault and rape: "the public nature of the living room meant gagging would spare the victims only from witnessing their own violation, but not from having others witness it".[107]

For Julian Dibbell, the author of this article, there are many soul-searching questions to be addressed. Not the least of them is "Where does the body and the mind begin?" After going through the possibilities, he has to conclude, "is not the mind a part of the body?" And doesn't that mean that mind abuse must have the same status as body abuse?

> Since rape can occur without any physical pain or damage, I found myself reasoning, then it must be classed as a crime against the mind – more intimately and deeply hurtful, to be sure, than cross-burnings, wolf whistles and virtual rape, but undeniably located on the same conceptual continuum.[108]

There are rape survivors who have not been subjected to severe physical pain but for whom rape is no less traumatising. (All women who are raped feel that their life is in danger.) There are women who have survived physical assault and abuse, but for whom the mental pain has never ceased. As a society we can accept that, years after the real-life violation has passed, the threat, the trauma and the pain that went with it can be as present and acute as they were at the time that it took place.

When we know that some things can be "all in the mind", we also know that as much mental pain and damage can be inflicted in cyberspace as in real life – and that we have to come to terms with a new category of criminal behaviour. The information revolution has brought with it the reality of mind crimes. Crime, punishment, codes of behaviour and social policy in cyberspace will all be major areas of concern in the future.

Already there have been some well-publicised examples of on-line sexual harassment and stalking that have resulted in legal action being taken. But as yet there are no national or international agreements on how to define mind crime – or even whether the laws of the land apply to cyberspace. One example[109] shows what can happen and indicates how the law might respond:

> Andy Archambeau wooed the woman of his dreams by e-mail, sending her computer messages saying, "Good morning beautiful" and invitations to rendezvous in the Bahamas.
>
> Now Archambeau has been charged with breaking Michigan's anti-stalking law for continuing to send computer messages after the woman and the police told him to stop.

Some of the free-speechers of cyberspace – along with the American Civil Liberties Union – considered defending Andy Archambeau on the grounds that the woman was

---

107. Julian Dibbell, 1994, "Data Rape; A Tale of Torture and Terrorism On-line", *Good Weekend*, 19 February, p. 34. "Gagging": not looking, switching off.

108. Dibbell, "Data Rape", pp. 37, 38.

109. The following account is taken from "E-mail Lands Amorous Man in Trouble", *Champaign Urbana News Gazette*, 27 May 1994, p. C1.

never physically threatened. But there were also computer experts who came forward to say that there was no difference between stalking someone physically and electronically. The woman herself didn't think that a virtual threat was any less real than a real-life one:

> He makes it sound like he was being romantic. But it was very very spooky . . . I wasn't sure if they were going to find me in a ditch. I didn't know what this guy was capable of . . .

It was a high-tech relationship gone wrong. The couple had met through a video dating service and found they had much in common – their computer habit included. For a while they corresponded via computer until she became uneasy and decided to end the relationship. And that's when he started threatening and stalking:

> The woman got worried when Archambeau left a message on her answering machine, telling her he had secretly watched her leave work. She filed a police report. Police told Archambeau to have no more communication with her, e-mail or otherwise. But Archambeau sent her another message anyway . . . After that he threatened to e-mail their story to all her computer friends on America Online and mail it to her family and old boyfriends. "This letter is the *least* of the many things I could do to annoy you," he wrote on April 24. Archambeau was arrested and charged on May 4. If convicted of the misdemeanour offence, he could be jailed for one year or fined $1000.

The idea that there can be crimes of the mind, on-line, still seems a strange one to many people, and it will take society some time to adjust. But this is nothing new. A code of acceptable usage had to be developed for other communication media and this is just another step along the way.

There will be protests, of course, particularly from the free-speechers. Already there has been such a case:

> A university student who transmitted a computer message describing how he wanted to abduct, rape and torture a female classmate claims he was merely exercising his right to free speech . . . "Torture is foreplay, rape is romance, snuff is climax" read one of the messages which identified the female student by name . . .
>
> "Just thinking about it any more doesn't do the trick. I need to do it." said another of the chilling computer entries.[110]

There is a battle being waged; it's a real one. If we don't want a repeat of the gender bias in the old law of the land, we have to insist that women are equally represented in the development of the new laws of cyberspace.

Some models exist. Laws had to be passed to prevent people from sending obscene or offensive material through the post, and it is illegal to make threatening, harassing or obscene telephone calls. But while there are similarities between the older forms and the new, there are also differences.

---

110. Peter Atkinson, 1995, "Court Told of On-line Lust", *Sunday Mail*, 12 February, p. 46.

The letter and the telephone call are, of course, part of virtual reality in their own right, in that neither of them is a face-to-face interaction. But still, they are not the same as cyberspace.

At this stage, we have not developed all the conceptual communication tools (to keep pace with the technological ones) so we have not got clear or refined ways of describing, analysing, or naming the nature of the leap into cyberspace. Commentators have tried to sum up the extraordinary difference between what we have been used to as human beings, and the potential of the electronic media in the future to shape our consciousness and our lives. Julian Dibbell has tried to describe the essence of it:

> the commands you type into a computer are a kind of speech that doesn't so much com-
> municate as *make things happen*, directly and ineluctably, the same way pulling a trigger
> does . . . [111]

You might not be able to "see" cyberspace, but it is no airy-fairy land, no paradise. It's a human construction, with all the human strengths and weaknesses. As Julian Dibbell has tried to explain, actions in cyberspace lead to real consequences. Just as in the real world, "for every action there is an equal and opposite reaction".[112]

This is why priority must be given to dealing with the endemic sexual harassment in cyberspace. Current practices are *not* about a few testosteroid males flexing their macho muscles and doing no real harm. Sexual harassment is becoming the modus operandi of the new world; the medium of communication for the generation of wealth and the support of power. It is the means by which some males are conquering and claiming the new territory as their own. This spells disaster for the hopes of an egalitarian ethos in this anarchic new world.

One of the starting points for change has to be in the educational arena. For more than a century women have been engaged in a battle for equal educational rights, and the struggle must now be transferred to the virtual society. We cannot continue to rest on our laurels and to believe that the battle has been won because women now achieve comparable results to men *in print-based systems and assessments*. Print is not where it is at any more – which is why women cannot afford to stay there.

But in moving on to tackle the new area where women must gain entry, there is one cardinal rule that has to be observed. It is that the place must be made safe for women. This is why Cheris Kramarae and H. Jeanie Taylor have set their priority as the formulation of a sexual harassment policy for cyberspace in all educational institutions.

> *A grievance procedure for complaints of sexual harassment on the nets.* This should be part
> of the sexual harassment policy of any institution that sponsors (or allows) electronic net-
> work communication. Users should be told that sexual harassment will not be tolerated, the
> prohibited behavior should be defined and punishments should be spelled out. Users should
> be given details on how, when, and where they can complain about harassment.[113]

---

111. Dibbell, "Data Rape", p. 38.
112. For further discussion, see Newton.
113. Kramarae & Taylor, "Women and Men on Electronic Networks", p. 56.

This section started with the recognition that in real life, girls are socialised from an early age to be streetwise. Increasingly, they are being trained in the art of self-defence. But in the opening paragraphs there was also the realisation that at this stage, we have no tradition of handing on advice from one generation to the next when it comes to protecting yourself as a female in cyberspace. The territory is too new. The wealth of experience that is needed for handing on has not been accumulated as yet.

But there are signs that this is changing.

At a "girls and computer" workshop in the United States, I watched one woman computer-whiz reassure the girls; they were reluctant, they didn't want to "play", and they were wary of "all the sexist shit and threats" they would encounter if they were to log on.

But the computer-whiz said, matter-of-factly, that this wasn't a problem. If they were interested, she would teach the girls how to program a "dog", which would "bark" and scare away all the unwanted advances that might come along.

The girls were very interested. Indeed, they were highly excited at the prospect. They worked in groups (their preference). In the space of a day they became proficient in this particular art of cyber-self-defence.

The transformation was extraordinary.

By the second day, these computer-reticent girls would have qualified as addicts. They were joyous. They roared with laughter when they set the "dog" off to bark at sexually harassing and unwanted messages. They wanted to learn all sorts of new possibilities. Some said that, while the dog bark was OK, they would like to go further and learn how to program a dog that could bite. Or worse!

It didn't take rigorous intellectual effort to work out what was at the root of this amazing change in the girls. Before they were taught to program for self-defence, they had felt too unsafe, too vulnerable, to log on. But after they had taken charge of their own environment – once they had even a few simple means of protecting their own space – they were more than happy to venture out.

Once they knew how it worked – that they wouldn't break it and it wouldn't bite them – the girls were in their element. Once in the position to counter the sexual harassment, they soon demonstrated they were perfectly capable of making their own way.

Not that I think self-defence is the ideal solution. As in real life, women who defend themselves can promote further attack. There is no doubt in my mind that the better women get at programming to counter the sexual harassment of cyberspace, the more intentional and insidious it will become. But until we can stop the men who engage in such behaviour, the only viable strategy is to ensure that women are not without a means of defence.

To the minimal requirements of a grievance policy, I would add the demand that all girls should be taught how to live, work, play and defend themselves in cyberspace. This comes to the nub of the problem. Because being computer-proficient isn't just about learning the mechanics of driving on the information superhighway. It isn't just the acquisition of a technological skill, even though this is how it is too often represented. To communicate via the computer is to become a member of a new society. Human growth and development skills are required along with the technological proficiency.

This is where women have a great contribution to make.

It is to be regretted that some of the worst aspects of real life have become prominent features of the virtual world. But women have a history of admirable achievements when it come to establishing a more egalitarian society. What we now face is a bigger task in our efforts to change the world: because now that we have cyberspace, we have a bigger world and more to change.

Virtual Dominance: Pornware

The distance from sexual harassment to pornography is often not great; indeed, they can be the same thing. If sexual harassment is the medium of on-line communication, pornography is the driving force for expansion. According to David Burgess, there are those in the pornography business who are not at all shy about declaring their cyberspace aims:

> The pornographers insist sex will be the service that spearheads acceptance of the new technology – the bait that lures the buyer. In the process, new high-tech gadgets such as CD-ROMs become standard household appliances. That's the theory.[114]

Using sex to sell a product to men isn't new, of course: for decades cars were sold by scantily clad females whose role it was to seduce the male buyer. But there is a much more direct link between the new technologies and the availability of women's bodies – real or otherwise.

It's because they can rely upon the demand for pornography that the purveyors are so cocksure of their prosperous future. Where pornography goes, many men will follow. A captive market, no less. With every other unsavoury and belittling practice making the transfer to the cyberworld, the only question that arises is why it has taken the professionals so long to muscle in on the area.

> "It's a huge market and largely untapped," says vice-president of technology and information services for *Penthouse*'s parent corporation, General Media. "It's the first really radical change that our company has faced in its 25-year history. Interactive erotica will secure our future."[115]

One reason that the professionals are making a grab for their share of cyberspace is that pornography on the networks is already widespread. While at the moment it is being peddled mainly by amateurs, this is about to change as professionals enter the marketplace. It's not as if they have to get out and sell the product, either; the buyers

114. David Burgess, 1993, "Hardcore Software", *The Australian*, 1 June, p. 9.
115. Burgess, "Hardcore Software", p. 9.

are already there and waiting. Faced with the opportunity to transmit pornographic images and to play with the new interactive media, it seems that boys make use of the toys simply because they are available.

What has to be kept in mind is that "Anyone with a computer and a modem can write anything he or she wants and put it on the net for anyone else to browse at any time." Millions can be reached by one posting. "It is therefore, a pornographer's dream."[116]

Even those who would not buy or trade pornographic print materials are not averse to plugging into the electronic forms. Or to passing them on to see what effect they have on the recipient.

Countless examples could be quoted of schoolboys who are experimenting with transmitting obscene and offensive images. Teachers acknowledge that these same students would not bring *Penthouse* into the classroom, they would not put pin-ups on the wall, and would not necessarily harass female classmates in real life. But they appear to have no compunction about making, storing, and transmitting these offensive images to the screens of the girls in the virtual world. And they seem to positively relish the discomfort that results.

Likewise, Lynda Davies has provided a chilling description of the computer science lab.[117] So prevalent is the porn on campus, and so awful the scenario that she outlines, that one despairs of ever being able to control or eradicate this offensive practice. In some universities – in the United Kingdom for instance – legal measures have been used to try to prevent faculty and students from misusing the resources of the educational institution.

> Vice Squad officers swooped on the Department of Metallurgy at Birmingham University and discovered thousands of pictures stored in the computer system of youngsters engaged in obscene acts. The material could be accessed through the Internet communications system to academic institutions, government departments and corporate business across the world. A 25-year-old research associate at the university has been interviewed by police in connection with the network.[118]

Universities are by no means alone in having their systems rorted. It seems that the business world and the office are not far behind: "A computer consultant recently stated in an interview that every large office he had ever been to had pornographic images stored somewhere in the computer system."[119] This is the general consensus of those in the industry who will talk about it: "In the video games industry, a variety of interactive, erotic computer games known as 'pornware' are finding their way on to corporate computer networks."[120]

---

116. Johnstone, "Surfin' for Sex", p. 9.
117. See pp. 182–3.
118. "Computer Child Porn Ring Busted", *Age*, 16 April 1994, p. 21.
119. Dianne Butterworth, 1993, "Wanking in Cyberspace", *Trouble and Strife*, Winter, p. 33.
120. John Hiscock, 1993, "The Chips Get Down to Sex", *Herald-Sun*, 2 October, p. 22.

It is interesting to note that, while an extraordinary amount of pornography is being sent around, not a great deal of public fuss is being made about what it means and who gets to view it. There is the occasional article in which protests are registered, but in general, there are no vocal organisations or international campaigns that are attracting media attention.

Even on campus the tendency has been to play down the situation. Cheris Kramarae and Jana Kramer, for example, report on one university in the USA where pornography was freely pushed to women students, and where the women involved not only complained, but put forward a series of recommendations to deal with the offensive behaviour. But "the matter was never included on the agenda at a faculty meeting".[121] Whether this was because the topic was considered too big, or too trivial, is difficult to determine.

Much of the material that currently exists and is shifted around has been made in the name of "boys will be boys". We are asked to believe that it's all about fun and games. While the validity of this excuse can be in doubt, there's no doubt about the demand for sex and sleaze. It's everywhere on the net.

There can be no doubt either about the money that can be made.

A recent *Cook Report* on ITV looked at bulletin boards specialising in computer-based pornography. One computer consultant was charging £350.00 for helping men to access over 10,000 hard-core pornographic images and to set up their own pornographic bulletin board. It was only a matter of time before the commercial pornography producers realised how much revenue they were losing . . . the largest producer of pornography in Germany estimates that within five years the majority of the porn they produce and distribute will be in video format.[122]

Many adults who are not computer-proficient, who don't do on-line communication, have no idea about the extent of the pornographic material available and how easy it is to get. If they have children who spend time in cyberspace, it's simply not possible for them to monitor their children's participation. Whereas parents can keep a check on *Penthouse* or *Playboy* (or the even worse publications that appear in print), and can insist that certain programs on television be switched off, there is no ready way they can implement an "appropriate-use" policy on the superhighway.

That kids can get hold of this pornographic material without any great difficulty is a point made dramatically by an investigation team with the *Sydney Morning Herald*:

As part of a two-month investigation into the use of computer bulletin boards, the *Herald* easily obtained hard-core pornographic pictures, stories, games and full motion video animations via telephone calls to a number of Sydney bulletin boards.
   Among the materials were:
• Video depictions of oral sex, intercourse and masturbation.
• Written stories detailing incest, rape, child sexual assault and bestiality.

---

121. Cheris Kramarae & Jana Kramer, 1995, "Net Gains, Net Losses", *Women's Review of Books*, February, pp. 33–5.
122. Butterworth, "Wanking in Cyberspace", p. 34.

- Still and full motion video descriptions of explicit sexual acts and of naked children.
- Sexually explicit computer games.

One Sydney bulletin board offers users a menu of images: digitised photographs and stories detailing acts including sexual intercourse, homosexual acts, incest, child pornography and bestiality. No proof of age or identity check is required.

Another offers digitised pictures described as: "Topless blonde covered in oil"; "Filth at its best! Dykes"; "Two girls and some lucky guy"; "Big-busted blond all tied up with ropes"; "Naked woman getting whipped"; "Some bitch getting her ass caned"; and "Two teenage girls doing it together".

Simply by dialling STD, the computer user can access a mass of information from nearly 1000 regulated bulletin boards across Australia alone.[123]

With the generation gap in computer competency, it is more likely to be the younger than the older male members of the community who at this stage are calling up the pornography. We must not underestimate the incentive that interactive pornography can offer; as the commercial providers indicate, it will spur many to buy the latest equipment and to work out quickly how to use it. At the moment, however, "Teenagers are the biggest market for this pornography," according to David Killick and Sonya Sandham. "These services are readily available and there is very little checking of age." They continue:

The extreme youth of some computer porn enthusiasts surprises Mr Platt (system operator of the massive 1990 Multiline bulletin board), who coordinates bulletin board directory listings for Western Australia.

"A recent listing was put in by an 11-year-old," he revealed. "And where you list the various options available, one of them was 'adult areas'."

. . . "As soon as I saw that, it stood out like a red rag to a bull. So I rang his mother. And his mother was, to say the least, quite distressed at the whole thing. She didn't have the faintest idea what he was up to."[124]

Writing in the *Courier Mail*, Craig Johnstone reaches the same conclusions. He describes the many porn sites (including a run-down on the brothels in the world, Australian ones among them) and says that it took him less than ten minutes to locate these sites. For anyone who is connected, there is a map available.

There is another newsgroup in Usenet that is far more graphic than the "basic" sex newsgroup. It contains descriptions of violent sex, rape, incest and paedophilia. It, too, receives hundreds of new postings daily and is about the most demonstrative proof available that we should be extremely concerned about some of the things of which the Internet is capable.

To describe what this newsgroup contains as pornography goes only part of the way to explaining the nature of its contents. Like most other newsgroups it is unrestricted.

All of this would be inconsequential if it were not for the net's availability to children and its potential as an educational resource.[125]

123. David Killick & Sonya Sandham, 1993, "Policing the Techno-nasties", *Sydney Morning Herald, Spectrum*, 18 December, p. 1A.
124. Killick & Sandham, "Policing the Techno-nasties", p. 1A.
125. Johnstone, "Surfin' for Sex", p. 9.

That the most obscene pornography can be so readily accessed by young people on-line is not in dispute any more. The only matters now open to debate are: what effect does this have on those who are exposed to it; and, as a society, what do we want to do about it?

I do not need to be convinced that pornography has a significant impact on the psyche – and the behaviour – of the human beings who make use of it. There is a growing body of research which sensibly makes the connections between pornographic and violent videos and games, and sexually exploitative and violent thoughts and behaviour.

Abusive and rapacious messages take their toll on the individual – and the younger they are, the more impressionable:

A youth from Long Island described how ecstatic he was about the long awaited release of *Mortal Kombat*, the goriest, most explicitly violent video game ever to go on sale. The move he really liked, the youth said, was "punching" an enemy off a ledge and watching him become impaled on a spike. Near the end of the game, the young player specially relished the choice between being able to electrocute his opponent and then either rip out his still beating heart or tear off his head and, like a gladiator, hold it aloft for the crowd, its spinal cord dangling from its neck.

Acclaim Entertainment Inc., the US company that makes the game, expects to sell two million copies in America and elsewhere. In a year-long trial run in America's video arcades, *Mortal Kombat* was number one. This is an especially depressing thought at a time when teenage violence everywhere is increasing at alarming rates.[126]

And the "target" for such violent abuse is female, more often than not.

I subscribe to the public safety and accountability theory when it comes to pornography.[127] It is, in essence, that in our society, manufacturers cannot put a product on the market which might put public safety at risk. All mind-altering and behaviour-changing drugs, for example, have to be rigorously tested for any risks to health before they can be dispensed. If, after they have been issued, there is a threat to public health or safety, the manufacturer can be sued. Likewise, the onus is on the producers of electrical and other goods to ensure that there is no danger to society in the products they manufacture. And the same rules in relation to health and safety should apply to the porn producers.

The burden of proof should be on them to establish that porn does *not* corrupt, disturb, destroy the values of the individual, or put at risk the life or limbs of any member of the community. Only when they can prove that their product has been properly tested and found to be perfectly safe for human consumption, should they be allowed to sell their wares.

This would make an extraordinary difference to the nature of the material available.

For some strange reason, however, pornography has been seen, not as a product,

---

126. Peter Pringle, 1993, "Video Games Link Bloody Fantasy to Violent Reality", *Age*, 22 September, p. 15.

127. My thanks to Janet Irwin and Beryl Holmes for this explanation.

but as a form of free speech. This is a remarkable quirk of the classification system. It has meant that those who want pornography do not have to get their hands dirty by defending sordid sadism and sleaze. Instead they can adopt a much loftier position and argue for its existence under the banner of free speech.

Because free speech occupies the high moral ground, it's harder for the average person to be against it, than to denounce the savagery of pornography.

You can see how this works in practice.

Several thousand racist and sexist jokes were put on a bulletin board at the University of Waterloo, Ontario, and made available to other institutions. Some universities decided that the transmission of such material was a misuse of their facilities, and decided to keep it out of the system. Stanford University was one of them.

Because the information was considered offensive and incompatible with the aims of the university (which are "to provide an educational atmosphere free of racism, bigotry and other forms of prejudice"),[128] a decision was made at Stanford to prevent the "jokes" from being accessed; anyone who tried to call them up found the following message:

> . . . Jokes based on . . . stereotypes perpetuate sexism, racism and intolerance . . . This bulletin board does not serve a university education purpose; its content is offensive; it does not provide a forum for the examination and discussion of intolerance, an exchange of views or the expression of views of the members of the University community . . .[129]

But one professor of computer science was categorically opposed to this decision. In the name of free speech, he copied the "funny" material on to a university research computer so that staff and students could access the information whenever they so desired.

There is little that can be done about this. It is made possible by the mantra-like defence of free speech, although what sexist and racist slurs – and the production of representations that are subjected to violent abuse – have to do with free speech is a matter for speculation.

We take for granted the fact that you can't just say what you like about people; no one is free to slander, libel, threaten, defame or abuse another individual. So why should there not be laws against threatening, defaming, demeaning, or abusing a gender, or a race?

It's partly because the legal status of cyberspace is so uncertain that there are difficulties about framing laws for the control of pornography. "The nature of the law is reflective and responsive, not pro-active," comment Cheris Kramarae and Jana Kramer.

> The law reflects the status quo: that is, it looks to history, precedent, tradition – it asks "what was done before?" But in cyberspace nothing has been done before – it is a whole new entity.[130]

---

128. See Kramarae & Kramer, "Net Gains, Net Losses", p. 33.
129. Kramarae & Kramer, "Net Gains, Net Losses", p. 33.
130. Kramarae & Kramer, "Net Gains, Net Losses", p. 33.

That the law has not yet caught up with the technology is a point that is frequently made. Trying to keep track of what is happening on the superhighway is extremely difficult for law-makers, as the Australian Attorney-General explains:

"The problem with policing bulletin boards is defining what is publication," said Michael Lavarche. "It is a new territory legally. At what point does it become publication?"[131]

The laws which relate to publication are quite clear when it comes to pornography in print, but they will not transfer directly to the electronic medium. So, for example, it could be perfectly legitimate to read or view something on your screen, but if you downloaded or printed it, then you could be engaged in an act of illegal publication.

This could mean that anyone sending porn on-line could be quite within the law; and that only those who made a copy would be committing an infringement. If the point is to prevent the images being sent to people who don't want them, this is hardly an acceptable legal situation.

An attempt has been made in the United States to bring the technology and the law into line; legislation has been introduced to stamp out "digital pornography". Put forward by Senator Jim Exon (Democrat), the bill is intended "to prohibit the distribution of pornographic materials over public and private networks, and to prevent harassment through electronic mail." According to Senator Exon, it would stop the "information superhighway from resembling a red-light district."[132]

Carriers or service providers would be liable if they "knowingly" transmitted offensive material over their networks, and this worries some of them.

On-line service providers are wary of the proposal, particularly with increased use of the Internet.

"With 200 World Wide Web home pages created every day, there's no way an on-line service can control what's coming into the pipeline. We don't create the content," said Mr Brian Ek, communications director for Prodigy, an on-line information service.[133]

In Australia, some service providers are already trying to ensure that they are not held liable for what is sent through their system, or how it is used when it arrives on someone else's screen. One Brisbane-based bulletin board, Psychopolis, warns users: "Some of these files may be prohibited in hard copy form. It is *not* recommended that you print these files."[134]

In his article "Pornshops on the Internet Not Easily Closed Down", Frank Crawford[135] comments that no Parliament in Australia has gone so far as drafting a bill which might

131. Killick & Sandham, "Policing the Techno-nasties", p. 2A.
132. Jill Gambon, 1995, "Networks Wary of Anti-porn Bill", *The Australian*, 28 February, p. 27.
133. Gambon, "Networks Wary of Anti-porn Bill", p. 27.
134. Killick & Sandham, "Policing the Techno-nasties", p. 2A.
135. Frank Crawford, 1995, "Pornshops on the Internet Not Easily Closed Down", *The Australian*, 28 March, p. 39.

address these issues, although politicians are, of course, very concerned. They are likely to be even more concerned once schools are connected to the Internet as the norm. While there may be a degree of tolerance associated with the computer science professors and students transmitting and receiving porn over the Internet, the idea that school students should be engaged in the same activities would lead to an angry protest. Yet there's not a lot that can be done to stop such material being generated, and published throughout cyberspace.

It just isn't possible to monitor or censor all that is "out there". Nor is it possible in many cases to trace the originator of the obscene products. Individuals who aren't doing it for money, and who don't want to be identified, can always find ways of concealing the source of the information. And, as the service providers have pointed out, it's not fair to hold them responsible for the behaviour of the users; just as it wouldn't be fair to hold road-makers responsible for the behaviour of dangerous drivers.

So you can't stop it from going on the net. You can't stop it from being transmitted. And because of the international nature of cyberspace, you can't even use the existing state and national laws for the purpose of prosecution. Even if you could overcome all these obstacles and find a technological way of policing the porn, which body could possibly be the censoring authority? Frank Crawford considers this question in relation to Australia.

> The only organisation legally empowered to do so is the Australian Film and Literature Classification Board.
> I doubt whether the board is even aware of the Internet. I'm positive it's not in a position to view and rate 50Mb of news a day, spread over more than 3000 newsgroups.
> The ultimate problem with censorship on the Internet is due to its international nature and the fact that everyone on the Net can act both as producer and consumer; it is therefore impossible to impose a policy that will cover everything.[136]

It is the shift from consumer to producer of porn that makes pornware so different. Printing files is one thing; having virtual sex is quite another. With all the interactive possibilities now available, it is fair to say that the superhighway has more in common with a red-light district than with an erotic bookshop.

The magic of interactive media is also the menace. On-line sex, in all its various guises, brings with it the possibility of prostitution and organised crime. No doubt law-makers will soon have to take this into consideration as they tackle regulation in cyberspace.

For anyone who has no direct experience of on-line sex, the following examples are provided. They are by no means an exhaustive list of what you can get in cyberspace; nor are they selected by any form of systematic sampling. But they do suggest some of the scenarios that can be called upon in virtual reality. To begin with the basics: "*Penthouse* now have a bulletin board called *Penthouse Online*, where men

---

136. Crawford, "Pornshops on the Internet Not Easily Closed down", p. 39.

can download porn files, 'chat' with the models, and discuss whatever it is that porn users discuss."[137]

This, however, is kids' stuff in comparison to the programs that are available for Mac-users:

> In MacPlaymate, one of the first on the market, the player summons an animated rendering of a woman, "Maxie", who flutters her eyelids, moves her mouth, breasts, legs and hands and entices the user into the program.
>
> The player can then strip Maxie, force her to engage in a variety of sex acts and employ any of six devices from a "toy box".
>
> For emergency use, the program includes a panic button that instantly transforms the pornographic image into a fake spreadsheet should the wife or boss approach.[138]

Not that Maxie has had a long lifespan. She has been displaced by a sexier player.

> MacPlaymate creator Mike Saenz has updated this version into Virtual Valerie, one of the best sellers in the CD-ROM market at 100,000 units. Virtual Valerie is a little tougher than her predecessor and if the user answers "no" to the question, "Do you want to go to the bedroom now?" she turns off the computer in disgust.[139]

It's the "realness" of the new porn which makes it so very different from anything that's been around before. As with every other type of information in cyberspace, the viewer/user becomes a participant; a "doer" of pornography rather than an observer of it. This is a point emphasised by Dianne Butterworth: "Interactive pornography means that the consumer is no longer just the consumer, he is in a sense, the producer as well."[140] This gives an added sense of power to the user and reinforces the powerlessness of the used.

> Interactive computer pornography is an alarming but logical progression in computer porn – the interactive exploitation of women. Not only does the male consumer look at the porn, he controls it. *Penthouse Interactive* holds 45 minutes of digitised video . . . The user plays the role of a *Penthouse* photographer; he chooses which one of three women to photograph, how much (or little) she wears (she strips off the required amount of clothing), and in which position she lies/sits/stands. The user takes a photograph, and Bob Guccione (the owner of *Penthouse*) judges how good (pornographic?!?) it is.[141]

And this is just "fun and games": it is simply beyond me to outline some of the stuff that is available on the information superhighway.

Although there are more and more individuals being connected to the net, it is still the case that most people don't know what it is like to do porn in the virtual world.

---

137. Butterworth, "Wanking in Cyberspace", p. 35.
138. Hiscock, "The Chips Get down to Sex", p. 22.
139. Burgess, "Hardcore Software", p. 9.
140. Butterworth, "Wanking in Cyberspace", p. 35.
141. Butterworth, "Wanking in Cyberspace", p. 35.

Which is why there are some graphic accounts for the uninitiated, in newspapers:

> The voluptuous female security guard meets you at the entrance to the beach-front apartment block. You enter her office and begin to spy on the sexual activities of the residents through the internal monitoring system. You choose the scenes that interest you most ...
>
> The guard is becoming aroused. She asks you if she can remove some of her clothes; if she can touch herself. Eventually her boss arrives. Excited rather than angry, he begins to have sex with the guard, stopping only to ask you what they should do, and how. Oral sex, spanking, straight intercourse? All of the above?
>
> How do you respond? Easy. Just press the key on your computer that corresponds to your preferences and the actors on your screen will perform as directed. Don't worry. There are no limits. And next time you can choose something different.
>
> Welcome to computer sex. Welcome to the 90s.[142]

There is still considerable debate as to whether these games are pornographic, or whether they are but an extension of "normal practice" to cyberspace. "Computer sex. Depending on who you believe it's either the technological renaissance of the sexual revolution – leading to enhanced communication and libidinal understanding – or just more electronic sleaze."[143] I have no difficulty with my own verdict! Nor do the majority of women.

Those who seek to defend cybersex do so on a number of grounds. Free speech. Liberation. In these days of HIV-AIDS and sexually transmitted diseases, safe sex can be the rationale for making sex on-line available.

> Computer laboratories in California are now said to be close to perfecting the ultimate in "safe sex" – the virtual reality orgasm.
>
> Cyberspace requires only one participant and when the technology is perfected, scientists claim the results will be almost as good as the real thing.[144]

But if the "realness" of the activity gets the punters in, it is also the element that many women find most menacing and degrading. Bad enough that men should drool and fantasise over printed images and have their attitude to women shaped by the sleaze that can be found on many pages of men's magazines. But how much worse it is when the image of the woman can be controlled, subjugated, maimed:

> The director of communications for the National Coalition Against Pornography, Ms Nancy Clausen, says women are more degraded with the computer porn because they can be physically manipulated.
>
> "This is dangerous because it can lead to people acting out their fantasies, especially children, if they are exposed to it," she says.[145]

---

142. Burgess, "Hardcore Software", p. 9.
143. Burgess, "Hardcore Software", p. 9.
144. Hiscock, "The Chips Get down to Sex", p. 22.
145. Burgess, "Hardcore Software", p. 9.

As there are no physical or legal means of preventing this material from being put "out there", any protection against it will have to be undertaken at the receiving end. Software could be developed to filter material, in much the same way as control mechanisms have been proposed for children's television viewing, and stops have been put on international dialling on some telephones.

Craig Johnstone comments that, ironically, the demand is so great that it in itself is working as a regulator:

> Computer archives featuring nude or near-nude bodies are swamped by users as soon as they are put on-line. On its first day on-line earlier this month, *Penthouse*'s site was visited by more than 800,000 people. *Playboy Magazine*'s cyberspace edition gets about 600,000 visitors a day. . . .
>
> There is usually a queue into these sites. The waiting and downloading time to get Internet pornography on to a PC can be quite long. And with service providers charging about $9 an hour for the privilege of hooking into the net, time is money.[146]

Such a demand, of course, also spurs the technologists and commercial porn producers to come up with faster, better, bigger and greater means of supply.

Not much faith can be placed in the possibility that porn consumers won't be able to get what they want when they want it. Which is why Frank Crawford is of the opinion that the most important contribution to restricted access will probably have to come from parental guidance, and education.

> We need to educate people to not undertake such activities and for those who are still too immature to understand this, their parents or guardians need to guide their choices.
>
> Certainly it is necessary to set up certain enclaves that can be guaranteed to be clean, much as is done by controlling what is available in libraries.[147]

While I think this is a bit of wishful thinking – wishing for a better society and a different notion of masculinity – I nonetheless applaud the sentiment. There is clearly a need to do something to ensure that boys cannot purchase digital, interactive prostitution and porn simply by means of a keystroke.

What needs to be acknowledged is that all the arguments against porn in the real world apply to cyberspace with even greater force.

It's not just that such offensive material will be
- readily available;
- readily transmitted to those who don't want anything to do with it;
- almost impossible to prevent or police.

It is that interactive porn is of quite a different order from all the print – or film or video – forms that have been on the market to this point. And while the principle might be difficult to grasp, what has to be understood is that interactive porn allows men to *do* dominating, raping, violent sex, to women. And children.

---

146. Johnstone, "Surfin' for Sex", p. 9.
147. Crawford, "Pornshops on the Internet Not Easily Closed Down", p. 39.

This virtual behaviour will have real consequences in the virtual world. And in the real world as well.

What's good for the goose doesn't apply to the gander

Women are silly creatures who don't know how to use the Internet. Or so says Bernard Morley in his letter to the computer pages of *The Australian*.[148] All that women are interested in are "Crafts, sewing, knitting – most of the time," he says. Whereas men's concerns are much more serious business.

Here we have another fascinating example of the double standard.

Because men have been seen as superior in past centuries, a quaint myth persists. It is that whatever men are associated with is significant, important, prestigious, while whatever women do is petty and trivial. The famous American anthropologist Margaret Mead drew attention to this habit of having one rule which diminishes women and another which magnifies men when she stated in 1950 that

> Men may cook or weave or dress dolls or hunt humming-birds, but if such activities are appropriate occupations of men, then the whole society, men and women alike, votes them as important. When the same activities are performed by women, they are less important.[149]

The computer comes into this category. When women were the ones with the skills, it was "women's work" and what they were doing was regarded as mundane, mindless, and fairly unimportant. But these days it is men who monopolise the machine, and sitting at a keyboard has become "men's work"; what they are doing is now leading-edge and supremely powerful and significant.

It's easy to argue that it is the status of the sex, rather than value of the skills or the interests, that put men at the pinnacle. You only have to look at some of the examples of cyberspace to see that women are often doing constructive, creative things, while men are playing around, behaving in a juvenile, techno-tedious manner. But any attempt to call men's bluff – or to play by the same rules that work for them – can have some shocking consequences for females. As Janis Cortese found when she challenged the sexual double standard:

> As a long-time *Star-Trek* devotee, Janis Cortese was eager to be part of the Trekkie discussion group on the Internet. But when she first logged on, Cortese noticed that these fans of the final frontier devoted megabytes to such profound topics as whether Troi or Crusher had bigger breasts. In other words, the purveyors of this *Trek* drek were all *guys*. Undeterred, Cortese, a physicist at California's Loma Linda University, figured she'd add perspective to the electronic gathering-place with her own momentous questions. Why was the male cast racially diverse while almost all females were young, white and skinny? Then, she tossed in a few lustful thoughts about the male crew members.

148. April 1995, p. 26.
149. Margaret Mead, 1950, *Male and Female*, Gollancz, London; reprinted 1971, Penguin, Harmondsworth, Mddx, p. 151.

After those seemingly innocuous observations, "I was chased off the net by rabid hounds," recalls Cortese. Before she could say "Firephasers", the Trekkies had flooded her electronic mailbox with nasty messages – a practice called flaming.[150]

What emerges is that – just as in real life – it's OK for the males to see females as sex objects, but any female who questions the practice, or who tries to "reduce" a male by such stereotyped ranking, soon finds the wrath of the cyber-machos descending on her terminal.

Just as it is OK for men to mock and malign women on the net (for their supposedly pathetic interest in knitting and sewing, and their failure to acquire "the smallest amount of technological and scientific vision"),[151] it is certainly not on for women to respond in kind.

This message came through clearly when Susan Herring, Deborah Johnson and Tamra Di Benedetto did their analysis of conversational space on-line. Only once in a five-week period did the women have more space than the men. And the reaction was unambiguous. There were men who protested that they were being silenced by the women, and that the women were "man-bashing" bitches.

> We suggest that this reversal – the fact that women were talking more, and on a female-introduced topic – made men uncomfortable to the point of threatening to unsubscribe and that it was ultimately responsible for male perceptions of "silencing" and female dominance in the discussion.[152]

The men actually took up much more space than the women had ever had, in making their protest against this "guilt-tripping" and man-hating they were being subjected to. It goes to show that it is OK for men to take the floor, but it is a very different thing when women try to do so.

One of the most interesting ironies to emerge from the "gender-aggro" on the Internet is that it is often the same men who insist on their right to free speech, on the principle that anything goes, who are the first to fume, and to flame the women who dare to dispute the male monopoly.

When in 1993 *Computer Underground Digest* published an article by Mike Holderness, outlining the discriminatory treatment of women on the net, a certain Larry Landwehr let fly with a frenzied attack on such views as a threat to his free speech:

> If *CUD* truly believes in "electronic freedom", then it should stop publishing articles that lay the groundwork for censorship and governmental restrictions. Instead, it should use its editorial discretion to promote positively written articles that will benefit the net and lead to its further expansion into the mainstream of human [*read* macho male] culture.[153]

---

150. Barbara Kantrowitz, 1994, "Men, Women and Computers", *Newsweek*, 16 May, p. 48.
151. Michael Lines, 1995, letter, *The Australian*, 21 March, p. 24.
152. Herring, Johnson & DiBenedetto, "Participation in Electronic Discourse in a Feminist Field".
153. Larry Landwehr, 1993, "Some Comments on the *London Times Educational Supplement* article", *Computer Underground Digest*, April 22.

Very few researchers have even pointed out that free speech often amounts to free speech for the white man: that women and people of colour, for example, have always had to watch what they say. They have always had to see which way the wind is blowing before they can express an opinion.

Because they feel as though they do not have free speech in the presence of men, many women have set up women-only forums. (The same applies to other oppressed and minority groups, of course.) But where such safe space has been set up, the response from some quarters is predictable. There are men who vehemently object and who claim that women-only space impinges on their right to free speech. They try to over-run, disrupt or destroy the exclusively female forum.[154]

That women's freedom of speech is not the issue, but that it is men's that must not be threatened is another revealing example of the sexual double standard. It's a case of what's OK for the goose would never be tolerated by the gander. And the gander rules in cyberspace – unfortunately.

As Susan Herring *et al.* have pointed out, it is very difficult for women to be able to register a protest about the cyberspace double standard. When, for example, an appeal was made in *The Australian* for a fairer deal for females, Michael Lines responded with a scathing attack on women's stupid and pitiable state.[155] His stand was shared by other men who were prepared to go public with their loathsome rhetoric:

Michael Lines' brief and brilliant letter was confirmed rather than contradicted by the mountain of irate replies from feminists.

Instead of answering what he'd said, they waffled about looking after children (no one forces you to breed, lady: if it's such a hassle, don't do it). . . .

In hundreds of Australian companies, men's careers are being stuffed while women of ordinary ability and lesser experience are promoted to fill a cosy little quota system called affirmative action.

When I hear of plans to "take the feminist flame to cyberspace" I know how the people of Poland felt in the lead-up to Nazi invasion.

Feminists have already poisoned the universities and the workplace with their irrational hatreds and vindictive legislation.

Now the refuge created by men in cyberspace can expect the same feminist onslaught of censorship, fines, imprisonment, thought and speech control, and fear – all imposed predominantly at the expense of male tax payers.

(Ah if only the law allowed us to treat them the way they treat us: if only these bitches could just once, taste their own medicine!)

Women are welcome in cyberspace as long as they behave like decent human beings [*read* cyber-machos!] and don't violate its spirit of freedom and equity. But the feminist Gestapo has no right to control cyberspace or anything else.[156]

154. Wylie, "No Place for Women", p. 4.
155. *The Australian*, 14 February, 21 March 1995; see p. 201.
156. J. Jones, 1995, letter, *The Australian*, 11 April, p. 26.

It's so awful that you could be forgiven for suspecting that it is a parody. But you do so at your peril.

Were I to critique men in the same terms that this man has vilified women, I would face public condemnation and censure. Or worse. As Cheris Kramarae writes in her cyber-breaking article, "A Backstage Critique of Virtual Reality":

> If I made a program that portrayed men in the ways women are portrayed in these programs (such as *Virtual Valerie*), I'd likely be dead very shortly. Freedom of speech is only for a relatively few adults, almost none of whom are women who counter the "malestream" voices.[157]

Of course, not all men play by the macho rules. Some are prepared to stand by women who are seeking a fairer deal. But they can be regarded as traitors and can find themselves on the receiving end of the same vituperative and vicious denunciations. (This was what happened to the man who supported the women's protest in the Susan Herring *et al.* study, for example.)[158]

Mike Holderness found that there was a savage response when he raised the issue of one rule for both sexes on the net. The following is one example of the reception he got:

> Here's a word of advice for the women on the net: If you can't stand the heat, ladies, then get out of the kitchen.
>
> Stop whining about how unfair the world is. Stop hiding behind paternalistic (maternalistic) governmental laws. Stand on your own two feet and earn some respect!
>
> Sexual harassment on the net, with no possibility of physical contact, is nothing but another type of flame. Learn to handle it. Learn to give as good as you get.
>
> Forget the fact that men enjoy technology because they like gadgets and naturally gravitate to the net. Forget the fact that women are late comers to this and many other fields. Forget the fact that men are naturally adventurous and are usually in the forefront of exploration. Forget all these logical reasons. Let's just say that men are oppressive. Let's not talk about paying your dues and taking your knocks until you manage to ensconce yourself on the net. Let's not talk about getting a thick skin so you don't get blown away by the first flame that's directed at you. Let's blame those rotten, bad, insensitive men instead.
>
> Women of the net, conduct yourselves professionally, and over time you will get the respect you want and will then deserve.[159]

It could almost be said that when women raise their heads above the parapets and become visible, there are men who are prepared to flame them down. The unpalatable fact is that it is men who have the clout and the weapons. They can make sure that everything is played their way. Margie Wylie quotes one particular example where Goliath deprives Davida of the smallest possible space:

---

157. Cheris Kramarae, 1994, "A Backstage Critique of Virtual Reality", in Steven G. Jones (ed.), *Cybersociety: Computer-mediated Communication and Community*, Sage, Newbury Park, Calif., p. 48.

158. See pp. 193–4.

159. Landwehr, "Some Comments on the *London Times Educational Supplement* article".

When an on-line service dedicated to women's concerns, Women's Information Resource and Equity (WIRE for short) was forming, *Wired* magazine immediately threatened the small service with swift and punitive legal action. The glossy, gossipy "journal" of Internet culture claimed that WIRE was too close to its own name for comfort, yet the magazine has gone on to register <macdonalds.com> as a domain name on the Internet for no other apparent reason than to taunt the Infobahn-impaired corporation. A legal tangle with McDonalds is a funny, nose-thumbing prank: a small women's service one letter removed from your own name is a dire threat, a matter of self-preservation. *Wired* defended its territory in the best tradition of the culture it celebrates: it flamed. In the end *Wired* demanded and got a name change. Now the service goes by Women's WIRE.[160]

To some extent, there has always been one rule for those in power and one for those without – so it shouldn't come as a surprise to find that the cyber-machos have got it all worked out They have written the rules in their own interest, and it is more than clear that many of them don't intend to see any changes in this convenient arrangement. "Organise your resistance now!" is J. Jones' injunction to his fellow males who don't want women as women on their turf and who are prepared to go to extreme harassing lengths to keep them out. [161]

The double standard that exists in the real world will be transported to the cyber-world as well. This process is already under way. We find that the male values and interests – including those of being top dog, and of having the most say – are now in operation. The best prospect is that they are not sufficiently entrenched to be beyond the realm of woman's challenge, and change.

It is possible that, in the future, there will be but one rule for women and men: indeed, one rule for all diverse groups on this potentially egalitarian network. Yet it won't necessarily be that men's rules prevail. It is relatively easy to make a case that *what's good for the goose is better for the gander*. That's the hope for cyberspace.

Women's Way

Girls Who Can

There is no doubt that women can be the computer networkers of the future. They can be as much at home on the net as they are on the telephone. You only have to look at the girls at Methodist Ladies' College, "the ladies with the laptops" (Chapter 5), to know that in the right circumstances, girls can make the leap to become producers and consumers of the new technologies. They can bring to this mind-shaking, mind-making medium some of the issues, interests and investments that are women's way, and which women also bring to real-life society.

Because there are as yet not many women (relatively speaking) on the net, this book has gone into some detail on the barriers to women's participation. (For example, the

---

160. Wylie, "No Place for Women", p. 5.
161. Jones, letter.

nature of the software, the open-endedness of cyberspace, the content of the CD-ROMs and games that are on offer, not to mention the sexual harassment and pornware that discourage women's involvement.) But while it is clear that there are obstacles placed in women's way (including those of cost), it is also the case that women who have "got through", who are committed computer networkers, speak in deliriously positive terms about their experience. (There are days when I am one of them.)

There's not much point in finding out what turns girls off computers, unless we also start to think about what turns them on. Who are the women who are the computer-whizzes? How did they get to be computer geeks? And what does cyberspace look like to them from the inside?

These are the women who are being catered for in the new on-line and hard-copy publication *geekgirl*. It is a "quarterly zine for geek gals and aspiring geek gals",[162] published by Rosie Cross and Lisa Pears. Rosie Cross explains the name as a pun on the popular image of the computer geek:

> Geeks generally are the hardcore of the nets. They are usually self-taught, determined individuals who simply love computers – machines are not just their friends but conduits into the world of art, politics, fun, magic and mayhem.
> Being a girl I am *voilà* – geek girl.[163]

The zine contains varied accounts from women about how and why they became computer-committed. Joell Abbott is one of them.

> As I sit pounding away for hours on a keyboard, it is hard for me to believe I once considered myself computer phobic. . . . I yelled loudly whenever *anything* went amiss . . . I also knew that if I should touch the wrong key, some bit of infinite wisdom would be lost to the world forever, so I kept a safe distance between computers and myself.[164]

These are the fears and the feelings of apprehension that many women report when required to form a relationship with a computer. Many stop at this point; they give up and decide that they don't want to take on the trials and tribulations of the technology. Joell Abbott was one whose attitude changed: from a computer-phobe she turned into a computer addict. She gets "withdrawal symptoms until the e-mail is read each morning". As she sagely says, "There are worse fates."

It was a Mac that got her in. It was the way it let her draw a straight line. "I am a very frustrated artist type," she writes. "I have no drawing talent, but with the Mac I began to do graphic arts for fun."

That was the start of it: the enjoyment, the creativity, the activity, the entry to a new way of life. And – as many women report – Joell Abbott found the Mac highly compatible: "Another great thing about Mac is that we speak the same language."[165]

---

162. *geekgirl*, 1: STICK, p. 2.
163. Quoted in Rodney Chester, 1995, "A Place in Cyberspace", *Courier Mail*, 13 April, p. 22.
164. Joell Abbott, 1995, "Computer Phobic Tale", *geekgirl*, 1, p. 3.
165. Abbott, "Computer Phobic Tale", p. 3.

Such testimony helps to show the importance of the software – the interface.[166] Joell Abbott found the Mac inviting: she found the drawing/art capacity enticing and satisfying. And she was away.

So for anyone who wants to interest girls in computers, Joell Abbott's experience could serve as Lesson Number One. Make it inviting and exciting, stimulating and rewarding. The sad fact to be faced, however, is that this is not the way most young women are introduced to the new technologies. There aren't many for whom it has been a positive affair from the start.

While interested in the specific personal experiences of women and computers, Dr Sadie Plant (from Birmingham University, UK) is also very concerned with the general relationship between women and the information infrastructure. She deliberately refers to herself as a *cyber-feminist* to draw attention to the connection between women and the machine.

> To start with I simply used the word cyberfeminism to indicate an alliance. A connection. Then I started research on the history of feminism and the history of technology. It occurred to me that a long-standing relationship was evident between information technology and women's liberation. You can almost map them on to each other in the whole history of modernity. Just as machines get more intelligent, women get more liberated.[167]

This is indeed a refreshingly new framework for evaluating women's experience and for thinking about women's entry to cyberspace. Sadie Plant continues in the same vein; she banishes some of the "mythic barriers" that have been erected to keep women out of cyber-territory.

According to Sadie Plant, women should feel comfortable in cyberspace. After all, when you stop to think about it, the medium is more attuned to women's way of working in the world than to men's. Cyberspace has the potential to be egalitarian, to bring everyone into a network arrangement. It has the capacity to create community; to provide untold opportunities for communication, exchange, and keeping in touch. In other words, it is like an enhanced telephone.

As far as Sadie Plant is concerned, this means that it is men who are going to have to make the modifications; it's the female rather than the male style that will be the foundation stone of cyberspace. As she says, "really the notion that it is all masculine is a convenient myth sustained by the present power structures".

Sadie Plant's message for women is – don't be bluffed. Don't be put off. Don't believe the bumf that this new area of wealth, power and sheer enjoyment is not for you. It's the men who should be rethinking their position and acquiring some new attitudes if they want to feel at ease in the "uncontrolled" conditions of cyberspace.

---

166. See also Ben Shneiderman, 1989, "Designing the User Interface", in Tom Forester (ed.), *Computers in the Human Context*, Basil Blackwell, Oxford, pp. 166–73.
167. Rosie Cross, 1995, "Cyberfeminism: An Interview with Sadie Plant", *geekgirl*, 1, p. 4.

While she doesn't state explicitly that masculist values are dysfunctional, she shows that "might is right" is an unhelpful attribute in an intellectually demanding cooperative community.

But if she believes that girls already – have got what it takes – she also acknowledges that they must be given a more positive and rewarding initiation experience with the new technologies in the future.

> Generally girls are brought up to avoid interaction with technology. Nevertheless, women's relationship with machines is more intimate historically than is men's. Now, for instance, girls grow up with technology. It isn't new to them.
> Technophobia is increasingly becoming a myth. I think it is a shame a lot of feminist theory buys into the notion of technophobia. It not only buys into it, it's keen to perpetuate it.[168]

It is true that there are schools of feminist thought which are opposed to technology. But it is also the case that they have contributed an invaluable critique of some of the sexist practices of male-dominated machines (reproductive technology is a good example of this). While I completely endorse Kranzberg's first law that "Technology is neither good nor bad, nor is it neutral",[169] I think it's crucial that feminists should now become centrally involved in the discourse concerning the information infrastructure. There's nothing to be gained by dismissing the new technologies from a distance.

Cyberspace is here to stay. The most responsible stand that feminists can take is to find out how it works and to become decision-makers in the process. Clearly, women have to put aside any socialised fears we have retained, and be prepared to get our hands dirty.

> ROSIE X: Do you agree with Donna Haraway that technology is a deadly game?
> SADIE: Only for the white guys.[170]

Sadie Plant isn't the only one who thinks that all girls need modems. Another self-confessed addict who believes technology has more solutions than problems is Jude Milhon – more commonly known as St Jude (the patron of lost causes). St Jude's advice to women is that they shouldn't be intimidated by the technology or by the boys' pranks and porn on the superhighway. After all, they can't *hurt* you, she counsels women; it isn't as if you are alone. Cyberspace is a public space and there are plenty of people around to see what's going on. While some might think her message harsh and unhelpful, her assertion that data rape is very different from real-life rape deserves attention:

---

168. Cross, "Cyberfeminism", p. 4.
169. See Melvin Kranzberg, 1989, "The Information Age", in Tom Forester (ed.), *Computers in the Human Context*, p. 30.
170. Cross, "Cyberfeminism", p. 4.

Keep in mind in cyberspace – *everyone* can hear you scream. There was a woman crying virtual rape on LambardaMoo. It's a game, lady. You lost. You could have teleported. Or changed into an Iron Maiden (the spiky kind) and crimped his dick. But by playing in this way you've *really* lost. Because the Moo's also a social place where you can meet people with real cultural differences – like Klansmen – and make them respect you as a woman, as a dyke, whatever . . .[171]

I must admit that, while I don't share St Jude's optimism (that "You may change their prejudices for ever"), I appreciate the spirit in which the advice is offered.

It is indisputable that there are women in cyberspace – a significant number of them – who have not only come to terms with the power ploys of the boys, but who are prepared to beat them at their own games and on what they see as their own turf. This has to be seen as one way of dealing with the problem.

It also has to be taken as evidence that girls *can do* computing, no matter the conditions – if they want to.

The fact that all the students at Methodist Ladies' College, Melbourne, are extremely computer-competent is confirmation that there's nothing wrong with the girls. The presence of so many geekgirls defies any myths or mutterings about girls lacking spatial ability, left-brain sophistication, or technological genes. Any barriers to girls' entry are strictly man-made, in the form of harassment, software and hardware.

One excellent example of girls who can do computing, and who are very much in control of the new medium, is VNS Matrix. These are the geekgirls who are challenging the machismo culture, subverting the masculist values. In their own words, they have "hijacked the toys from the boys" and have set out the "Cyberfeminist Manifesto for the 21st Century". They continue to work on their game, *All New Gen*, which is

> an interactive computer artwork and installation piece. . . . *All New Gen*'s mission as anarcho cyberterrorist is to undermine the "chromo-phallic patriarchal code" and sow the seeds of new world disorder in the databanks of Big Daddy Mainframe . . .[172]

The list of cyber-women who are trying to interest other women in the new medium is a long one. Spider Redgold, for example, has taken on the task of making it easier for women to use the net. She is the support manager at Pegasus (an Australian Internet server) and the women's net liaison person. She is a wonderful role model. As she says,

> This [the Internet] is going to be the mode of discourse for the 21st century. . . . If women stay with the telephone, which is too costly, and letter writing, which is too slow, we will be left behind and we will be left out of the discussion, decisions and conversations.[173]

---

171. Rosie Cross, 1995, "Modem Grrrl; Wicked St Jude", *geekgirl*, 1, p. 8.
172. "VNS Matrix", 1995, *geekgirl*, 1, p. 5; see pp. 188–9. The Australian Film Commission has granted $100,000 to VNS Matrix to develop their game.
173. Debbie Tucker, 1995, "Spider Spins Web for Women", *Courier Mail*, 13 April, p. 22.

Spider Redgold will be helping to network the UN Conference on Women in Beijing in 1995, so that women around the world will be able to access it on-line.

"Princess Internet" is the name that another cyber-woman, Katherine Phelps, is known by; she has published a book, *Surf's Up; Internet Australian Style*. Like many of the younger generation who have been reared in the electronic environment, she can't remember when she wasn't engaged with a computer. Her earliest memory is of "afternoons spent on her scientist grandfather's computer, happily lost in the endless electronic story game, *Colossal Caves*".[174]

Katherine Phelps practises what she preaches. Not only has she written a book about the joys of the Internet, she has also gone into business on the net. She has set up Glass Wings, one of Australia's most visited Web server sites.

She obviously believes that the Internet should be fun. She spent nine months out visiting on the net, sampling what is on offer, and selecting the most interesting and helpful places for readers and users. For women, the advice she gives has a reassuring ring: "You just have to get out there and goof around, push buttons and try it out. Things won't break!"[175]

All these competent women belie the belief that women can't do computing. Women like Karen Hellyer in the USA, who spends half her time as a high-school art teacher and the other half at the National Super-Computing Center, and who has established a students' art gallery on the Internet; Mary Hocks, who developed one of the first (and still one of the best) programs that integrated computers and creative writing; Maureen Ebben, who did her PhD on women's experience of net communication; Elizabeth Reid at RMIT who is an internationally recognised net surfer and researcher; and Cheris Kramarae who, with H. Jeanie Taylor, has done so much to develop research priorities in the area through WITS.[176] The existence of all these leading-edge women (and here I have named but a few) opens up a particular strand of research: what makes them tick?

Are they special, or is such competency an option for all women?

There are many related questions: How did these women come to computing? What barriers, if any, were there? How did they overcome them? And how do they feel about their relationship with technology – their connection to a machine?

These questions are not just a matter of interest; if we know what drew these women to the new technologies, we would have some idea about how to get more women on-line. This is one of the reasons that Sherry Turkle has been engaged in her invaluable research.

For more than twenty years, she has been looking at the relationship between computers and human beings. Besides this general and extensive background, she has also made a specific study of twenty-five Harvard and MIT women students who are geek-

---

174. Lisa Mitchell, 1995, "Princess Internet Takes Her Glass Wings to the Ball", *Age*, 7 March, p. 27.
175. Mitchell, "Princess Internet Takes Her Glass Wings to the Ball", p. 27.
176. WITS – Women, Information Technology, and Scholarship – at University of Illinois, Urbana Champaign.

girls or computer-whizzes. One of the interesting findings that emerges is that even these highly professional young women still have a slight sense of unease about their relationship to machines. "Lisa is 18, a first-year student at Harvard, and surprised to find herself an excellent computer programmer. Not only is it surprising but 'kind of scary'."

Sherry Turkle is *not* surprised by this response. She thinks that "women's phobic reactions to the machine are a transitional phenomenon":

> There is the legacy of a computer culture that has traditionally been dominated by images of competition, sports and violence. There are still computer operating systems that communicate to their users in terms of "killing" and "aborting" programs. These are things that have kept women fearful and far away from the machine. But these are things that are subject to change.[177]

Because the women at Harvard and MIT have grown up in this male-dominated computer culture, and because they have been socialised as girls to distrust the toys that are taken up by the boys, it is understandable that they should be reticent about the machine. But their reservations have not been sufficient to stop them from becoming engaged and enjoying the experience.

Nancy Kaplan and Eva Farrell are equally concerned with the ways women get hooked on computers, but they look more at the on-line environment than at the demands of computer programming (which was Sherry Turkle's starting point). Their subjects were five adolescent girls for whom electronic communication is a major leisure activity. (Eva Farrell is one of the adolescents as well as an author of the report on the group's behaviour.) The researchers ask: What do these young women have in common? What drew them to the net? And do they provide us with any insights into how we might lure more women into the electronic communication web?

One of the first points they address is the extent to which research on girls has concentrated on issues of tension and exclusion. They warn against this, because of the danger of constructing knowledge which emphasises the problem, which presents a "deficit model". Such knowledge can become a self-fulfilling prophecy. This is an issue regularly raised by geekgirls who – from their inside position – can get impatient with those who only see from the outside, and who tend to dwell on the barriers and obstacles in the way of their entry.

Nancy Kaplan and Eva Farrell also suggest that this preoccupation with the problems is a generational one; for the young women who have grown up in an electronic environment, many of the so-called difficulties seem to have disappeared. In their sample, they found electronic communication the norm. Eva Farrell's comments about her participation are both perceptive and provocative.

With one exception, these adolescent geekgirls had a computer and modem at home.

---

177. Turkle, "Computer Reticence", pp. 41, 42; see above, pp. 173–5.

While they attended different schools, the schools were all of the "alternative" variety; they all valued intellectual freedom and curiosity. So it is a pretty privileged bunch being put under the microscope in this study. Not that this is a criticism of the research; it's just a reminder that access and equity are crucial factors. What we want is for all students to have the opportunities and advantages of this group, or those enjoyed by the geekgirls at MLC.

In trying to determine what the group had in common, Nancy Kaplan and Eva Farrell consider the vexed role of mathematics. It is often said that, in order to be good at "computing", you have to know mathematics (and do computer science). The group quickly takes up this point, and suggests that it is not so simple:

> Fish [a nickname for one of the young women] intends to study science. Farrell prefers literature, or perhaps psychology. Fish takes computer classes and writes computer code; Farrell does not and shows no interest in programming. In fact, she describes herself as math-averse and yet shows an intense interest in computers and communication tools.[178]

Farrell's "deviation" is not at all remarkable. Mathematics – like Greek and Latin before it – has often been used as an entry requirement rather than as a necessary skill for a particular activity or occupation. I remember when it was widely held that you couldn't *do* English unless you knew Greek (or Old English, or Anglo-Saxon, etc.); or that you couldn't *do* medicine unless you knew Latin. So it is important that we actually examine the role that maths and computer science play in computer network competency. With women's recent history in relation to mathematics, it could be that the very idea that you have to be good at maths is enough to deter some women from experimenting with computers. In fact – as I can testify – maths has no more to do with on-line facility than it has to do with answering the telephone.

What counted much more than mathematics to these young women computer buffs was the fact that they were all introduced to the net by a friend.

They didn't get there by computer games: none of them does any "wargaming". What they all share is a love of the creative and communicative possibilities that the net provides. They all acknowledged that it was the "nattering" that got them in:

> The conversations among these young women and their contacts on the b-boards often seem, at least to an outsider, driven more by the desire of the participants to keep the conversation going than by their desire to achieve understanding or consensus about some topic or issue. Often the messages are quite short – almost like conversational rejoinders.[179]

It's clear that for these young women the net is an extension of their social world; they are using it – like the telephone – to keep in touch. (They contact each other during the day at their different schools.) But it also allows the opportunity to construct an ongoing text – a combination of a good book and a telephone:

---

178. Kaplan & Farrell, "Weavers of Webs".
179. Kaplan & Farrell, "Weavers of Webs".

The sociability of this exchange seems its sole reason for being. Even though the conversational partners seem to be engaged in a dialog carried on over several days, or even weeks, the exchange itself has some of the qualities of rapid repartee. A more extensive examination might in fact show that these conversants were engaged in . . . "rapport" talk, a style of conversation more common among women than men, rather than "report" talk, a style men tend to favour.[180]

Every time I hear someone criticising computers for the "alienation" they promote and the drop in standards they represent, I feel the urge to quote some of Eva Farrell's comments to them. If the next generation share Farrell's values, social skills and standards, there is not much to complain about:

Farrell's journals and descriptions of herself and her interactions with others on the network show that she loves the word – spoken and written. For her, "Net people are people in love with knowledge. In love with information and words. Debaters, jokers, storytellers, discussers, users of the paths insomnia carves, along with solitude in the wee hours." The sense of the human connection and its value to them emerges strongly in the words of both these young women [Fish and Farrell]. The net seems to extend their connectedness to others, to work for them precisely because it connects them to others.[181]

For these young women, the net is as much a part of their lives as the telephone. They meet friends on the net, and forge relationships; they know of people who have dated on the net, and been married on the net (although divorce on the net is not mentioned). Their work, leisure, and social life will increasingly be conducted on-line, and with no great loss to human communication, linguistic standards, or interpersonal skills. Indeed, it is not only clear that women can do the new technologies; it is precisely because of what they can do that all women should be able to participate in the electronic culture. As Eva Farrell summarises it:

The Web is a reality of constantly shifting virtual truths: Identity, language talk, programs, even a deadly virus or two, circling through these invisible pathways of the information jungle out there. It is a book forever being written, revised and erased: a world that is inside one dimension of text on a screen, and yet does not exist in physical space. Is it any wonder I feel a surge of power through me when I exercise even my pitiful skill?[182]

Women's Networks

I have never heard it said that a reason women like talking on the telephone is that it is one place where they don't have to put up with the interruptions of men. Yet such an explanation makes a lot of sense. Women's speech is generally restricted and restrained when men are around. This is why women need women-only places where they can say what they think without fear of harassment or intimidation. It's why

---

180. Kaplan & Farrell, "Weavers of Webs".
181. Kaplan & Farrell, "Weavers of Webs".
182. Kaplan & Farrell, "Weavers of Webs".

women need everything from women's studies to women's refuges in real life; and why they need women-only networks on-line.

> Separatism is not necessarily what women want for their discussions, but until gender equality becomes a reality in a wide variety of settings, women need both access to the public nets and some safe forums for their conversations. (In case some . . . believe that only those "radical feminists" want women-only forums, they should know that there are all kinds of women-only nets, including a fundamentalist-Christian board.)[183] On women-only nets, the talk is more candid, and the topics, unlike those on "regular" boards, are controlled by women.[184]

Setting up women-only networks is one thing; keeping them women-only is quite another. Despite the rigorous vetting that takes place, it's almost impossible to ensure that people really are what they claim to be on the information superhighway.

> internet-women-info
> This list is for *women only*, period. If your name is gender-neutral or culturally associated with a man's name, or if you sub using only initials or numbers, I may ask you to send me a brief note assuring me you are a woman (born or TS). If I ask you to do that and you do not respond, I will assume you are male and will unsub you. In the interest of fairness I ask that all men currently on the list unsubscribe immediately. You should join the help-net list instead. To subscribe, send an e-mail to: . . .[185]

More common than the strict women-only networks are those set up by women, with a women's agenda, which men can contribute to if they want. No doubt this arrangement has evolved because of the practical difficulties of keeping the space for women only. But it is also because women-only lists attract such hostile attention from some men who think they have a right to go anywhere and to say whatever they please.

The problem is that, as soon as some men get in on the act, it becomes much the same as in real life. Men do most of the talking (or posting) and correcting of women.

Cheris Kramarae and Maureen Ebben know what they – and other researchers – think about women-only talk in cyberspace. "The literature is clear," they state. "Men are more interested in having mixed-sex conversations than women are. Women are more likely to give men plenty of time, space and consideration than vice versa."[186]

While some men move into the areas women have set up because they are *inflamed* by the sheer fact of women's presence on the net, others move in because they like the support and feedback (some might say "ego-massaging") that the women supply. Either way, women usually have to struggle to avoid being marginalised and silenced:

---

183. Angela Gunn, 1991, "Computer Bulletin Boards Not Just Boy Toy", *New Directions for Women*, November–December, p. 7.
184. Kramarae & Taylor, "Women and Men on Electronic Networks", p. 57.
185. Internet communication, 19 March 1995.
186. Kramarae & Taylor, "Women and Men on Electronic Networks", p. 57.

Even open networks where the topic is women's issues are, we have found, over-run by men. For example, on this campus (UIUC), we've checked the network called soc.women and have noticed that on some days only men have posted messages. On a national soc.women board, dominated by comments from men, one woman noted that even the women-only boards are not respected by men who cross-post to the women from other boards.[187]

Unfortunately, there are so many men querying, correcting and attacking the terms and priorities of these women-organised bulletin boards that many women have complained that the climate is too hostile for them and have withdrawn as subscribers.[188]

Nevertheless, women have found ways and means of carving out their own space; sometimes they just use it until it gets too hot, and then move on. One of the first and most successful initiatives was the Electronic Salon; it was planned by Deborah Heath and hosted by Lewis and Clark College (USA) in conjunction with their Gender Symposium. There was nothing but positive feedback from the women who signed on – in the early stages.

The Electronic Salon (April 1992) . . . afforded a unique opportunity for academic women across the United States and the world to be electronically linked. Once connected to the salon by means of electronic mail, the more than 200 participants could download the invited research essays on gender and technoscience and discuss the papers . . . The Electronic Salon provided a rare forum for the sharing of ideas and interests about women and information technologies as the usual constraints of time, money and distance that frequently prevent women from attending an academic conference were significantly diminished. Moreover, because this electronic "gathering" was organized primarily by women, focused on women's issues, and had women as its main participants, it was potentially empowering. (However, there was a disparity between men's and women's participation rates. As the total number of messages from men increased during the week of the symposium, the participation of women decreased. Women appeared to be shut out of the discussion, which as time went on, was increasingly dominated by the men.)[189]

As Cheris Kramarae and Maureen Ebben point out: there's no doubt about what happens to mixed-sex discussions on the net. Unlike the telephone, where women who talk to women are free from the dominance of men, women who talk to women on-line are constantly threatened by male interruption and overwhelmtion. This is why women go to seemingly extraordinary lengths to set up women-only networks.

There are literally thousands of women's groups now on-line: everything from a women's Web Site and Women's Resources on the Internet, through to Jane Austen, Women's Health Hotline, The Ada Project (a collection of resources for women in computing) and a Women Artists' Archive. Everything you could dream of wanting.

187. Kramarae & Taylor, "Women and Men on Electronic Networks", p. 55.
188. Kramarae & Taylor, "Women and Men on Electronic Networks", p. 55.
189. Maureen Ebben & Cheris Kramarae, 1993, "Women and Information Technologies: Creating a Cyberspace of our Own", in Taylor, Kramarae & Ebben (eds), *Women, Information Technology, and Scholarship*, pp. 17–18.

And more "exchange" than books could ever provide, which is why it is so exciting and gratifying.

SYSTERS is one of the women's networks which is the most successful and longest surviving:

> SYSTERS is a private discussion group established exclusively for professional women in computer science. Items include pleas for career advice, how to handle difficult situations, sexual harassment and current research. SYSTERS provides a networking and mentoring service that addresses the isolation often faced by women in computer science. An encouraging sign pointing to the value of electronic networking is that the SYSTERS enrollment has grown so large that a SYSTERS-ACADEMIC group has split off to address the particular needs of women computer science faculty. To subscribe to the SYSTERS group, contact Anita Borg at e-mail address <systers-request@decwrl.dec.com>.[190]

There are many more women's groups on-line: some have estimated that there are now thousands. A large proportion of them are academically oriented (an indication of academic utilisation of the net), but there are lots of other types as well.

> FEMCON-L is a discussion list for feminist economists . . . FEMINIST, owned by the Feminist Task Force of the American Library Association, discusses issues such as sexism, racism, pornography, censorship and ethnic diversity in libraries and librarianship . . . FIST is an unmoderated list addressing issues of feminism and science and technology . . . GAYNET focuses on gay and lesbian concerns . . . GENDER is a moderated group devoted to questions and answers pertaining to the study of communication and gender . . .[191]

One of the latest is Virtual Sisterhood, which is described in the second issue of *geekgirl*:

> Virtual Sisterhood is a new electronic resource for women on-line. Though still in its infancy, it is rapidly developing an extensive network for women around the world to share information, advice and experiences. Resources include the electronic conference/mailing list: a WWW site for archive info on Virtual Sisterhood, women produced electronic resources, FAQ [frequently asked questions] and related info: and a consulting service for women on information strategies and technical assistance for developing electronic information resources. Currently under development is their directory/resource guide, a comprehensive list of women and organisations engaged in electronic communications and publishing, including service providers, trainers, information resource developers, activist groups, etc.

It might come as a surprise to find that "Virtual Sisterhood has recently engaged with networks of women in Senegal, Malaysia, Sri Lanka, Bangladesh, India, Brazil, Mexico, Argentina, Costa Rica, Uruguay, Chile, Peru, Tunisia, West Bank and Gaza, USA, Canada and the former Soviet Union Bloc . . ."[192]

---

190. Kathleen A. Turek & Judith Hudson, 1994, "The Future is Here", *Feminist Collections*, 5, 3, Spring, p. 13.
191. Turek & Hudson, "The Future is Here", p. 13.
192. "Virtual Sisterhood", *geekgirl*, 2, p. 14.

Increasingly, women are seeing the potential of the net and are using it to extend their community. This happened at the International Communication Association conference in Sydney in 1994. Women from all parts of the world got together and exchanged information; towards the end of the conference, there was some discussion about keeping in touch. The possibility of a newsletter was even suggested.

But then came the idea of an electronic network. That so many women at the conference were without e-mail addresses was in itself a spur to look at electronic communication and the issue of gender. (And, too, there was a need to think about the women from particular countries for whom access to the net was not a reasonable proposition in the foreseeable future.)

No sooner was the question raised than women from the UK, USA and Australia, for example, acknowledged that, while they were not connected to the Internet, most of the men on campus already had e-mail. It was agreed that this inequality should be taken up as a matter of urgency by the Feminist Caucus of the International Communication Association.

The consensus was that research and findings should be undertaken electronically. Arrangements were made so that anyone who signed on and who had no realistic hope of electronic access would receive a print-out. For all the women for whom e-mail was theoretically available, the pressure was on: get connected, as soon as possible. And if there are any problems, keep a record of them.

Offers of assistance were made by those who were already proficient, and WITCH was formed: Women In Touch; Communication and Help.

The conference was in July. The plan was to wait until November (to allow some women to get on-line and others to become competent) and then for all women who had access, in a two-week period, to post a narrative of their own experience of getting connected, a personal account of their entry to cyberspace.

Real space doesn't allow me to recount all these stories now. It is possible that they could form a database for future reference and investigation. It is crystal clear from the experiences that were related that for a number of the women, nattering on the net soon became as sociable, as pleasant, as reassuring and stimulating, as talking on the telephone. A women's network in cyberspace was soon flourishing.

Another US WITCH checks in. I have things in common with many of the other WITCHes who have reported.

I am basically a technological luddite who has only gradually been able to overcome fears similar to Roth's – fear of failing, fear of appearing dumb, not even knowing the right questions to ask, and feeling that sitting down for a while to explore hands-on and work at understanding the system would be helpful, but of course, that "spare time" never comes.

I started my education at the advanced age of 38, and got my e-mail account originally as part of a statistics course I took as an undergraduate. I was very fearful of "crashing the system" so did not use it much at the time.

As a graduate student I also reluctantly joined the e-mail system but have loved using it for certain things. I still haven't had that "sit down" or participated in any workshops or training sessions, but that will absolutely have to come soon.[193]

---

193. This and the following quotes are from the WITCH "salon", printed out December 1994.

Almost all the women mentioned that they had to overcome their fear that they would break something or crash the system:

Another US WITCH checks in.

No more lurking in the background.

My experience with the net is a combination of naive blundering around and knowing who to ask for help. Fortunately, our techies are quite accessible and usually helpful. We even have a help phone number for computer problems. I think phoning is ironic and typical. I do seem to spend a bit of time calling people when I am originally trying to link to them via e-mail or get into their library etc. . . .

My computer use started about 8 years ago . . . Mostly, we all just crunched our data . . . and used the box as an extra typewriter. . . . A few years passed and my friends elsewhere started to ask what my e-mail address was. Huh? was my usual astute reply.

. . . One day, serendipity arrived. I think her name was Laura, . . . the student assistant in the lab [who] was cruising the net. I asked her to give me an electronic tour and I soon had mail from lots of discussion groups . . .

Some days are more trying than others. The mainframe (my port to the world) has thrown me out twice and locked me out once in the process of writing this note. No need to get totally into conspiracy theories but it does seem odd . . . I do wonder how many novice users have the experience of being thrown out or locked out and never return. I also suspect that I may get quicker service because I do visit the overall faculty lab and have been around long enough to know many of the techies. This year I know I am getting quicker service because I'm an assoc dean . . .

Almost everyone was in awe of the system. They were unable to conceptualise how it works (what's in a microchip?) and sometimes at a loss to know how to impact on it to make improvements:

My first experience with e-mail was a terrible seminar. I came away with a few basic commands and a lot of fear – I was sure that I would do something to make the whole system crash!!! I have since learned that the system is quite resilient . . .

I have subscribed to a copy of on-line workshops on the Internet – Gopher and others. Of course as soon as I subscribed to the Gopher workshop, Gopher was removed from the system I work on. Users really need to have more input into the decisions that techies make concerning the types of services that are available to us!!!!!

I also feel a big time constraint. As an untenured assistant professor, I have so much to do that learning a whole new system and filtering through all that info seems daunting.

The frustration is acknowledged, but usually overcome.

Witches

I am sure you are wondering what I sent you in the last three pages of jumble. It is a great case study in itself. I spent one hour learning how to download from a disk. The result – this jumble. Am I depressed? Did I give up? Did I swear at anyone? Of course not!

It's all a learning experience. Ha!

Now to the research – by the way, I went home, ate lots of food, and am going to type the whole thing in because no-one . . . can tell me how to download! The Help line can't help.

One of the interesting findings to emerge was how many women mentioned the pleasure of nattering on the net. Despite all the fears and the frustration, there was still the recognition that it could all be worthwhile:

> It is clear that our sense of learning is intimately connected with pleasure. Many of these experiences demonstrate the way women are using this technology to achieve pleasure. Through networking and "net" playing. These examples can also provide insights into the way the computer has the capacity to be used for our own ends, both in terms of challenging and joining dominant power structures. It is also an opportunity to subvert the present private/public split. As a way to build networks, contribute to academic discussion . . . and even to maintain a gender-neutral identity to allow us to enter debate on an equal footing, it is clear that the "net" has a great deal to offer.

But. And I think this is ironic and important. Quite a few women reported being censured for using the net for "social purposes", just as women were admonished once for using the telephone for personal rather than business uses. It will be worth monitoring net behaviour to see if this trend develops in the future, because it is yet another way of discouraging women from utilising the net to network.

This could have crucial implications for the workplace, particularly for women who are support staff and who are logging into women's networks. "Rhia", a secretary in a computer science department, has been criticised by her boss and his peers for wasting computer time by "nattering" with women's on-line groups. The men clearly regard the games they play as business (porn included!) while Rhia's contacts with other women around the world are an abuse of the system.

Not that this detracts from the women's networks, or from the skills that women bring to them. It is more than likely that women's groups on the net will function in much the same way as the women's page of the newspaper or the women's room on campus; they will meet the needs of numerous women, and attract the unwanted attention of many men. And be the subject of further research.

Gender-bending

The objection that was raised when the telephone was first invented was that it wouldn't be possible to tell whether you were talking to a male or a female on the telephone; at least, that's what people were worried about. And there was a genuine basis for their concern. Because one of the first principles of conversation is that if you don't know what gender you are talking to, you don't know what to say, or where to look, or how to stand. And that's only the beginning of the communication.

Even today, there are topics that women would not ordinarily raise if there were men around; (and there are ways of looking and standing that aid communication among women, but which would be totally inappropriate if it was a conversation that you wanted with a man). And women wouldn't talk about some things in a particular way if men were within hearing. Everything alters – from terms to tone, from posture to demeanour and eye-contact – depending on whether women are talking to women, rather than to men.

In the 1970s when I was avidly documenting the conversation of the sexes, I set up several experiments to find out how women talked in the presence of men.[194] And I didn't always welcome the evidence because no matter how assertive and aware the women were – myself included – we invariably made men the centre of our linguistic attention, and showed them considerable deference. Which was pretty distressing for feisty feminists.

With cameras rolling, experimental situations would be set up in which one man joined (uninvited) a group of women who were engrossed in conversation. That the man would interrupt and the women would give him the floor was to be expected. But that the women would change the way they were sitting (and move to a more "closed" position), that they would arrange themselves with neck inclined, so that they were looking *up* to him rather than down, was a chilling reminder of the subtleties of socialisation, and its power to construct daily inequality.

In a memorable study of the way that we mould and modify our behaviour on the basis of gender, a group of feminist researchers undertook some observations on mothers and a baby. The baby was dressed in purple, and was unknown to the subjects. It was handed to a series of women who were told it was a girl. Almost all of them held the purple-clad baby fairly close and tightly, with their heads bent down, while they spoke in quiet, soothing, affectionate tones.

Then the women were told that a mistake had been made. That the baby was in fact a boy.

The change was dramatic. The baby was sometimes in danger of being dropped as each woman struggled to change the way she was holding it. Immediate attempts were made to move the "boy" further away, to jog him up and down, to speak more loudly and more robustly. And far from looking down on a male child, many women raised it to eye-level. More than a few went so far at one stage as to hold *above* them a three-month-old purple bundle that they believed to be male so that they could incline their neck and talk to the baby in the appropriate gender-deferential manner.

What was said also varied, depending on whether the baby was thought to be male or female. "Aren't you a pretty little thing" was fine when the purple bundle was taken for a girl. It was quickly replaced with "What a big boy you are", or some other such stereotyped comment, when it was taken for a boy.

How it was said changed dramatically as well. Whereas low-key and quieter tones were used when the child was classified as female, much more noise was made, and more action-oriented terms were used, when the women thought they were talking to a boy.

It is clear that the behaviour of the women was determined by what they thought was the right response to the different genders. It had nothing to do with the antics or attitudes of the purple baby.

---

194. See Dale Spender, 1980, *Man Made Language*, Routledge & Kegan Paul, London.

The extent to which this code is imprinted on our psyches can also be seen when in real life we have to talk to someone whose gender is unknown. We can't even make comments to a new-born baby – who has not a clue what we are saying – without having to ask, is it a girl or a boy? Only when we know this can we conduct anything like a conversation.

Anyone who has ever assumed that they were talking to one gender rather than the other is excruciatingly embarrassed if they discover they have made an error. Every-thing from body posture to vocabulary and tone is wrong if you think you are talking to a man, only to discover that it is a woman. Or vice versa. Instant changes have to be made so that behaviour is appropriate and you are not misinterpreted, or seen as insolent or inept.

This is why people who say they treat both sexes exactly the same are either very rude or very silly. Parents who say they rear their daughters and sons identically are quite self-deluded. All of our words, actions and body language are gender-loaded.

This is not just in relation to the spoken form; it applies to the written as well. In another classic study of the early 1970s, Philip Goldberg showed that the same words were rated differently depending on whether the author was said to be male or female. The same extract, be it on child-care or scientific method, was seen as more impressive, authoritative and substantive if believed to be the work of John Smith than if it was thought to be that of Joan Smith.[195]

Which suggests that you need a male name if you want to be taken seriously in print.

These are not the only factors that need to be taken into account. If there are pluses for man-talk, there are minuses for woman-talk.

> It is not surprising to find that there are no terms for man-talk that are equivalent to *chatter, natter, prattle, nag, bitch, whine* and of course *gossip*, and I am not so naive as to assume that this is because men do not engage in these activities. It is because when they do it it is called something different, something more flattering, and more appropriate to their place in the world.[196]

While there is negative feedback for women when they talk like women, it is no less negative (and is probably more so) if women break the code and speak in the same way as men. We are all aware of the double standard that allows the same words to be masterful if coming from a man, but are domineering, bossy, aggressive – even castrating – if they come from a woman.

While these issues are rarely raised in the context of free speech – and who possesses it – all this evidence serves to show just how deeply gender considerations influence what we can say and how we can say it. In real life – and in cyberspace. Although for men there may be advantages in some of the established practices (like having a defer-ential audience that rates your contribution as authoritative, regardless of content or

195. Philip Goldberg, 1974, "Are Women Prejudiced against Women?", in Judith Stacey *et al.* (eds), *And Jill Came Tumbling After: Sexism in American Education*, Dell Publishing, New York, pp. 37–42.
196. Spender, *Man Made Language*, p. 107.

evidence), for women the situation is grim. So you can see why women might find the prospect of gender-blindness attractive. They might find electronic networks, with their scope for gender neutrality and pseudonyms, an improvement on real life. Gender-blindness at least holds the possibility of being able to leave behind the baggage that goes with being rated as a female speaker.

The fact that the electronic networks are theoretically gender-, race- and class-blind, has certainly been put forward as a positive feature by those who want to encourage new net users. There are women who have adopted this gender-bending strategy, and who say how much easier they find it when their gender is unknown on-line. But this is only part of the story so far.

Just as we were not fooled by the telephone, we are not going to be fooled on the superhighway. (Almost no one makes a mistake about gender on the telephone. While we can't quite work out how we do it, and maybe we didn't do it well when the telephone was first invented until we learnt the new "language", it is nonetheless clear that we now pick up clues to gender identity on the telephone; and that we are accurate most of the time.)

We are doing much the same thing in the new medium. We are now working out ways to identify gender and race on-line. I suspect that at the moment, anyway, gender divisions are just too important in our life to let them go. We are starting to use virtual clues to make decisions about gender, as – in real life – we use the subtle clues of body posture, terms, tone, eye contact, etc.

In her paper, "'Ladies, Babes 'n' Bitches': The Genderless Exultation of Cyberspace", Nancy Kaplan shows how things are changing. Whereas a few years ago it was routine to praise the gender-neutral possibilities of cyberspace, she says, "A number of studies have begun to challenge the assumption that net personas lack all gendered behaviors." It's not the gender-bending and cross-dressing that go on in cyberspace that attract the attention, so much as the general communication habits. You can't get away from gender even in the most mundane interchanges. In looking at MOOs and MUDs, Nancy Kaplan says, "The practices suggest that even when personae try to construct genderless avatars, there is no social, communicative space without gendered speakers."[197]

"Farrell", one of the young women computer-whizzes (see pp. 233–5) gives us some idea of what is going on and what gender judgements are being made when she says:

> The person that I met the other day chatting has the alias, Armpit. I realised that as we were chatting, that I had the idea he was male, before I had any real reason to do so. I suppose part of it was the alias; very few females would pick the alias "Armpit". Also he had some ways of being that seemed more male; for example, he responded to my "Hello!" message with something like "Yes Madame?" Yes, that is part of my alias, but females don't generally respond that way.[198]

---

197. On-line information re conference program – abstract for conference.
198. Kaplan & Farrell, "Weavers of Webs".

At the same time, however, "Fish", one of Farrell's friends, shows that the element of anonymity that goes with electronic communication can also be helpful – particularly for females.

> Fish's journal sheds a little light on her use of this conversational form. Describing herself as a shy person "in real life", she writes that she "feels more comfortable typing out my feelings. My mind doesn't work quite fast enough that I feel comfortable in a normal conversation, but typing messages, I can express myself very well." Fish's account of her behaviour and her conversational style she and others employ may well yield additional insights into why some women like electronic environments: the absence of social cues (appearance, for example) and of immediately perceptible power differentials (gender, age and so on) create a more comfortable social space, Fish believes, for many people like her.[199]

The temptation to adopt a gender-neutral name – or even a male pseudonym – must be great, if it provides any protection from the sexual harassment that plagues so many women on the net. These days I assume that most women will take non-female names (or handles) simply to avoid some of the hassle. And that any blatant feminine names that are out there (such as "Marilyn"), probably belong to men in drag.

Of course, there have been other times and other circumstances when gender-bending was popular with women – although this was not the name it would then have been given.

In the eighteenth and nineteenth centuries, women who were audacious enough to write – to speak out and to be paid for their services – were likely to be condemned and criticised as not "proper women". As a result, the practice of writing anonymously or under a male pseudonym emerged. Women novelists such as Fanny Burney and Elizabeth Gaskell began their careers anonymously, in order to avoid the hassle, in much the same way as women adopt names like Fish and Dolphin on the net today.

Anonymity isn't reserved for women. Men can also "benefit". It has been pointed out that it is the very anonymity that allows some men to behave abominably on-line, in a way that they never would in real life. They feel powerful and protected by their pseudonym. According to Randye Jones, it is this anonymity, along with the ability to choose their gender, or any other identity, that accounts for the extremes of behaviour on the net:

> I credit the behavior to a couple of things. The first and probably the most influential is that anonymity and/or masquerade factor. On the net, you can be anyone you want to be – and when you divorce yourself from your real self, you are free to do/say more inflammatory things; after all, it's not "me" saying that; it's "the [person's alias]". The same sort of behavior results at costume parties or masquerade balls – when it's Mardi Gras for instance.[200]

To Rosie Cross, however, it is not simply that men use gender-bending to harass, while women use it to hide: "Generally, men, and some women have invaded the net

---

199. Kaplan & Farrell, "Weavers of Webs".
200. Randye Jones, 1993, "Netters Behavior and Flame Wars", on-line communication, September.

with a duplicity of personae and a Jekyll and Hyde attention-deficit disorder. Huff and bluff [and] I'll blow your cyberspace down."[201]

To return to the women writers, however, and the parallels with today: they didn't adopt male pseudonyms just to avoid social censure. It was because they wanted to be taken seriously that they gave themselves men's names. They didn't want to be dismissed as women, but to be treated with the deference due to men. So Charlotte Brontë wrote as Currer Bell, Maryanne Evans as George Eliot, and Ethel Florence Richardson as Henry Handel Richardson, in order to get a fair hearing. And some women choose men's names for the same reasons when they enter cyberspace today. Not much has changed from one century to another – nor from one medium to the next.

Despite the similarities with the past, there are new pressures at work today. Among them is the way that the electronic media break down traditional divisions. We can see how national boundaries have been undermined, how even subject divisions are being dislodged. It could be that rigid divisions of gender are also being weakened by this tendency to get rid of the old certainties.

No one could dispute that there are gender changes going on as women move away from their traditional private and oppressed role. A new relationship is being forged between the sexes; and while it is by no means equal at this stage, there is no doubt about which direction it is heading.

Women no longer feel obliged to stay within the confines of the male-servicing, ornamental role. There is more to life than being an attraction for men. On women we see sensible shoes, the ubiquitous black outfit, not to mention shoulder pads, dinner jackets, career suits and men's suits. We even have a caption for a picture of the singer k.d. lang in "feminine" garb which states: "You know times are changing when women begin dressing up as gay men who dress up as women."[202] This is gender-bending and, as the caption says, it isn't the sole domain of females.

At the same time that women seem sufficiently unfettered to move away from the blatantly feminine image of earlier days, it has become obvious that there are men who feel free to adopt some the props of femininity. This is not just a matter of men being involved in child-care (although this is an important development); it's that some men now "dress up" as women once did.

Sadie Plant interprets this as a departure from the old system, which divided the world into male and female, and then insisted that they were unequal categories. While I don't entirely endorse her point of view, it has interesting elements. When asked was it possible to escape gender on the net, she gave the following response.

> I think it's increasingly advantageous to be female. So many men take on female personae that the gender issue has become an increasingly murky thing to discuss. There is everything to play for. It's fascinating that men want to play at being women. It's an opportunity they

---

201. Quoted in Rodney Chester, 1995, "A Place in Cyberspace", *Courier Mail*, 13 April, p. 22.
202. Victoria Starr, 1994, *k.d. lang: All You Get is Me*, HarperCollins, London, between pp. 128 and 129.

have not had presented to them in the past. It also implies a recognition by men that to be a woman in the past was a liability, but now it's a distinct advantage and a privilege. The male is basically becoming redundant.[203]

This is gender-bending taken to the limit; it will have enormous implications for cyberspace – and real life.

---

203. Cross, "Cyberfeminism", p. 4.

Conclusion

Social Policy for Cyberspace

Access and Equity

Cyberspace might be a virtual creation, but it is a reality that is here to stay. And my attitude towards it is pragmatic.

Given that I have to learn to live with the cyberworld, I want the best possible outcome that can be realised. This means that I want to be involved – along with countless others – in the decision-making process of shaping the information infrastructure. I want national forums set up, public discussions organised, working parties created to determine priorities. I want some indication that there are plans to use the technology to improve the quality of life for all human beings; I don't want to see it used (or misused) to enhance the lives of the few at the expense of the many.

At the moment, most of the forums in which cyber-policy is being made are exclusionary. White, professional, English/American-speaking males have got the floor: and they are focusing primarily on technological issues – or pornography, property, and privacy problems. (A survey of *Wired* indicates that these are the hot topics.) It is easier to talk about the latest "toys" and to defend the concept of "free speech" for the boys, than it is to address the major social and political questions which go with the new technologies; it is the human factor that now demands attention.

No matter the source, some of the relevant data are already available. Despite the ideal potential of the new technologies to create a global, egalitarian community, a virtual world without barriers or divisions, the scene down on the ground is strikingly different. In ways we can appreciate and ways we do not yet fully comprehend, the real people in the real world are being divided up into the information-rich and poor: the "master minds" and those who are "kept in the dark".

In the United States, the wealthiest nation on earth, "many working-class children have little access to computers in or outside the classroom: for many children [in the

USA] and other countries, life is about just trying to live".[1] Even with the projected reduction in the price of hardware, the cost of becoming connected remains just too great for many average families in the First World.

Technology enthusiasts might see the present personal computer as another form of the Model-T Ford, and predict that there will soon be one in every home. But without extraordinary measures this is not a likely scenario. There are households still without television: and even more households without telephones. Even in the USA (not to mention countries other than the richest one on earth), there is a significant percentage of people without a home.

Another issue which must be considered is that the more advanced the technology is in the First World, the more difficult it will be to live, work, and communicate in the Third World, which has no such resources. And the further the Third World will be removed from the centres of wealth and influence.

Yet the solution is not so simple as giving computers to the poor (a proposal that enjoyed a brief spin from the leader of the US Republican Party in the House of Representatives).

In countries where children are dying of starvation, where there is little or no health care and no clean water, it borders on the obscene to talk about the pressing need for information infrastructure. It might even be awkward, if not impossibly difficult, in life-threatening circumstances, to convince the poor of the desirability or necessity of such a technological solution. It could be taken as yet another version of Marie Antoinette's derisory contribution: "Let them eat cake!"

The Indigenous people of Australia are a case in point. With 20 per cent of Aboriginal children in the Top End suffering from malnutrition, and living in an environment where they have no decent housing, sanitation or health provision, the recommendation that everyone should have a computer could appear as more a sign of contempt than a social justice policy.

This does not mean that Aboriginal people should *not* have computers; if computers are the medium of the future, then there is every reason for every member of the Indigenous community to have their own.

But this assumes a country in which a government wants its citizens to be connected to the information superhighway. Not all nations will be in this category. There are going to be authoritarian regimes (such as China) where one of the last things that officialdom will see as desirable is for the population to have ready access to on-line information.

Chinese scholars and students are using the Net to send messages which the authorities are powerless to restrict . . . "There's no way you can technically prevent individuals from accessing the Internet," the Beijing representative for the US telecom company Sprint, Mr Ben Chan, says. "But China can set rates that are prohibitive for individual users."[2]

---

1. Cheris Kramarae, 1995, "A Backstage Critique of Virtual Reality", in Steven G. Jones (ed.), *Cybersociety*, Sage, Newbury Park, Calif., p. 53.
2. Paul Harrington (in Hong Kong), 1995, "Cybersurfers Enjoy Anarchy", *The Australian*, 11 April, p. 37.

WITS women attending the UN Conference for Women in Beijing (1995) have drafted a recommendation which they want the UN to adopt. It is to make access to information a human right. This will help to put access and equity in cyberspace on the international agenda. As an extension of the aim of universal literacy, universal "informacy" could be readily accommodated by UN policy. It could provide a platform for assessing the implications of the international division between the information-rich and the information-poor.

The possibility of the global *village*, where everyone can have a say, is as yet a long way off. Currently we have the technology enthusiasts who declare that the Internet will be coming soon to every village in the world, at a price that all can afford. But the social and political reality is very different. With so many international, national, local – and familial[3] – inequalities in existence, the cyber-sociologists' consensus is that there will be an even bigger gap between the information haves and have-nots.

Even in First World countries, there is much talk these days about a growing under-class, a social group which is cut off from the benefits of society. The homeless, the hungry, the unemployed, uninsured, unregistered – and *uninformed*. These are the people who are not "in the know", who have no means of participating in the public life of the community. In some respects, the situation is not so different from the industrial revolution, where another underclass was created. Bitter divisions and class struggles developed; it took violence, campaigns, trade union organisation, and an evolving system of regulations to provide a semblance of social justice and harmony.

Today, when so much energy is being focused on looking forward, it might also pay to take a look backwards. Those who don't know their history are liable to repeat the mistakes of the past. (The fate of history itself is discussed a little later in this chapter.)

The social and political consequences now demand more attention than the technological advances. We don't have to give up the ideal of the global village, the electronic community where everyone can be consulted. But we have to do much more to realise the possibility. Priority needs to be given to research and policies which actually examine and deal with the impact that the new technologies are having on human beings, globally and nationally and locally.

Cyber-communities

What social policies now need is a cyber-dimension. It is not just inequalities that they need to address. The relationship between cyber-society and the real world also calls for examination. Real selves and virtual selves will be a feature of the new society, along with real and virtual communities and forms of organisation.

(Philosophers will have a field day. Trying to tell the difference between what is

---

3. Even with the television, the most powerful figure in the household generally gets to choose the programs; and when it comes to computers, it is the male more than the female who gets to monopolise the machine. To the point where one of the excuses put forward by the girls of MLC Melbourne for not doing their homework was that their brothers had taken their laptops and wouldn't let the girls use them.

virtual and what is real will be big business, particularly when so many virtual actions will have real consequences.)

Communication and information are central to the newly created cyber-society, and we have to look more closely at what it will mean to be a well-informed and civic-minded individual in the circumstances. Conditions such as information anxiety and overload could even become the starting points for human existence in the cyber-community. The range of information activities that will be the daily reality borders on the overwhelming.

> A globe-spanning web of computer networks offers millions of users the opportunity to exchange electronic mail, transfer files, search databases, and retrieve information from remote libraries, take part in real-time conferences, run software on distant computers. and participate in discussion groups on topics from autism education to yacht design.[4]

Just how much time will be spent on information acquisition? What will this mean in relation to work, family, and leisure activities?

There is no doubt that new communities will be established as people meet with one another in cyberspace, and swap and share information. But what sort of communities will they be? Will they be additional to the ones we have in the real world – or is it possible that the cyber-communities will become our real and primary communities? What consequences will this have for human understanding and organisation?

We can look back and see some of the changes that took place as society was transformed from a manuscript culture to a print-based one. The questions are: "what sort of a culture are we getting, and what sort of culture might we want, in the cyber-era?

Howard Rheingold, the guru of virtual reality, raises some of the fascinating issues about what we are, and who are our friends, in this new cultural milieu: Are "relationships and commitments as we know them even possible in a place where identities are fluid?" he asks. At this stage, in cyberspace everyone is guessing. That's not the only problem when it comes to sorting out who is who, and how we should relate: "On top of the technology-imposed constraints, we who populate cyberspace deliberately experiment with fracturing traditional notions of identity by living as multiple, simultaneous personae in different virtual neighborhoods."[5]

For as long as human beings have been asking what it means to be human, the issues of identity and community have been at the forefront of discussion. These are not the only vexed questions which will have to be faced as we work out social policy for cyberspace. Our major concerns in cyber-society – as in real life – will be in relation to exploitation, crime and abuse.

---

4. Margaret L. McLaughlin, Kerry C. Osborne & Christine B. Smith, 1995, "Standards of Conduct on Usenet", in Jones (ed.), *Cybersociety*, p. 90.
5. Howard Rheingold, 1993, "A Slice of Life in My Virtual Community", in L. M. Harasim (ed.), *Global Networks: Computers and International Communication*, MIT Press, Cambridge, Mass., p. 61.

Cybercrime

When it comes to cybercrime there are the obvious instances of pornography, sexual harassment and stalking, as we saw in Chapter 7. Then, too, there is the vastly expanding area associated with intellectual property; how to be paid for creative work and how to protect it from piracy in all its many and possible forms on the net.

There is also increasing anxiety about security: computer and credit-card fraud, for example. In his article "Good Cop, Bad Hacker", Bruce Sterling (the author of the best-selling book, *Hacker Crackdown*) takes up many of the issues of cybercrime and the new law-enforcement agencies that have been created to patrol the superhighway. This brings with it many of the ongoing concerns of a democratic society. According to Bruce Sterling,

> there's way, *way* too much blather going on about teenage computer intruders, and nowhere near enough coverage of computer cops. Computer cops are at least a hundred times more interesting than sneaky teenagers with kodes and kards . . .[6]

The potential for the computer cops to become a form of thought police has to be taken into account; it should be a key item on the cyber-policy agenda. There are social critics who have suggested that one of the reasons that totalitarian states no longer needed secret services like the KGB and the Stasi was because communities can now so easily and extensively be kept under surveillance electronically.

As David Lyon has pointed out in his book *The Electronic Eye*, there's little that isn't known about an individual – and a household – in the electronic world. From what you purchase at the supermarket, to where you were on a given day and what financial transactions you engaged in, right through to your credit-card rating and who you call on your telephone – all this information is available, and matchable.

Apart from the fact that every individual is vulnerable (because this information can be sold, stolen, stored and used in an exploitative way), there is also the issue of the invasion of privacy.

What people regard as private varies from one culture to another, and doesn't necessarily stay the same from one generation to the next. But there does seem to be general consensus that there should be a division between public and private, and that what is considered private should be protected from prying eyes. David Lyon asks:

> what exactly is threatened by the rapid rise of computerised record-keeping either by state or economic institutions? In Victorian times, the fear was that members of the public might obtain unseemly access to the private lives of elite people, such as politicians or the rich. British Royalty, among others, continue to struggle with this. With electronic surveillance, however, the equation is reversed. It is the lives of ordinary citizens that are thought to be at risk from large and powerful agencies. Indeed the practice of computer matching . . . tends to place the poor, the vulnerable, the minority, at a particular disadvantage relative to big bureaucratic forces.[7]

---

6. Bruce Sterling, 1995, "Good Cop, Bad Hacker", *Wired*, May, p. 122.
7. David Lyon, 1994, *The Electronic Eye: The Rise of a Surveillance Society*, University of Minnesota, Minneapolis, p. 14.

The greatest privacy at the moment, however, seems to be that surrounding the arrangements that governments are making for data management. David Lyon sees the new technologies as leading to "the rise of surveillance society", and is determined to point out some of the dangers.

This list of possibilities is by no means exhaustive; rather, it shows that cybercrime and exploitation come with the territory. As with every other public area where deals are done, wealth is made, and information is exchanged, there is a need for a code of appropriate behaviour. Policies which establish protocols and suitable sanctions for offenders are now relevant and necessary .

Whereas regulations, guarantees and forms of consumer protection have been developed in the past for manufacturing products (so that you can sue if the electrical goods are faulty, or the car is a lemon) we will now have to develop codes of practice in relation to software manufacturers, for example. If the plane crashes as a result of computer failure, or the medical drip machine dispenses the wrong drug because of a software glitch, does the computer programmer get charged with negligence, or even manslaughter? What are the individual, corporate and bureaucratic responsibilities in the virtual world?

It is hard to draw the line in some areas, but much easier in others. This is certainly the case in relation to acts of net terrorism. The mind boggles to think of the havoc and horrors that could readily be perpetrated on the net.

Just one example serves to illustrate the damage potential in this area.

Cynthia Sulaiman was the moderator of an Islamic group on the net, and the trouble started on 17 January 1995 when she turned on her computer and found that her password didn't work.

> "At first I thought it was a glitch," she said. . . . [but then] The next day mysterious postings started appearing from her account. In desperation she sent electronic mail from another account to warn of the hijacking. "Anything posted in my name [she wrote in the message] is actually due to intruders into my account, those who wish to do my reputation harm and regain despotic control . . ."[8]

It is estimated that 30,000 people round the world belong to this Internet group, and it is thought that two computer scientists were trying to oust Ms Sulaiman from her position and take control. She was sexually harassed and warned off by the two men. Her "resignation", asking for forgiveness, was also posted to the group members.

> Ms Sulaiman denies she ever resigned. But by now subscribers to the news group were getting confused. Who was the real Cynthia Sulaiman? And what were they supposed to forgive her for?

By last month, Ms Sulaiman had had enough. She was frightened of her adversaries – "I

---

8. Martin Bright, 1995, "On the Internet: The First Holy War", *Sydney Morning Herald*, 11 March, p. 23 (from *The Guardian*).

didn't know what they were capable of" – so she called in the FBI, who told her what was happening was a clear breach of the law.[9]

There will be many more breaches of the law, breaches that require new laws, as we rethink our ideas of what constitutes a crime. And how we are acutely information vulnerable in an information culture.

The Personal Becomes Mechanical

One of the best personal accounts of vulnerability and the way we are changing our values and understandings as human beings has been written by John Seabrook, a journalist with the *New Yorker*. One Friday morning, the unsuspecting Mr Seabrook logged onto his computer, only to find a flaming e-mail awaiting him.

> Crave THIS, asshole.
> Listen you toadying dipshit scumbag . . . remove your head from your rectum long enough to look around and notice that real reporters don't fawn over their subjects, pretend that their subjects are making some sort of special contact with them, or worse, curry favor, by TELLING their subjects how great the ass-licking profile is going to turn out and then brag in print about doing it.
> Forward this to Mom. Copy Tina and tell her the mag is fast turning to compost. One good worm deserves another.[10]

He was shocked by this. "I rocked back in my chair," he writes. After he had read it, he felt cold, distressed, shocked and shaky. He got up and walked around, but the message stayed on his screen. Gradually, the power that the new technology had over average people began to sink in. His identity as a white, professional man was under threat.

> No one had ever said something like this to me before, and no one *could* have said this to me before: in any other medium, these words would be, literally, unspeakable. The guy couldn't have said this to me on the phone, because I would have hung up and not answered if the phone rang again, and he couldn't have said it to my face, because I wouldn't have let him finish. If this had happened to me on the street, I could have used my status as a physically large male to threaten the person, but in the on-line world my size didn't matter. I suppose the guy could have written me a nasty letter: he probably wouldn't have used the word "rectum" though, and he probably wouldn't have mailed the letter; he would have thought twice while he was addressing the envelope. But the nature of e-mail is that you don't think twice. You write and send.[11]

This awful experience (which continued to haunt this hardened journalist) prompted John Seabrook to look more closely at communication on the net. As he points out, it is the very egalitarian qualities of cyberspace (once you get connected) which are both

---

9. Bright, "On the Internet: The First Holy War".
10. John Seabrook, 1994, "My First Flame", *New Yorker*, 6 June, p. 70.
11. Seabrook, "My First Flame", p. 71.

the benefit and the drawback. For the "same anonymity that allows the twelve-year-old access to the professor, allows a paedophile access to the twelve-year-old" he writes. And the "same lack of inhibition that allows a woman to speak up in on-line meetings allows a man to ask the woman whether she's wearing any underwear".[12] In other words, it is a medium for good or ill – it's the human factor that makes the difference.

The abusive human behaviour behind the flame that John Seabrook received made a very big difference to his peace of mind.

As he debated whether to respond in kind, he went so far as to itemise in his head all the vilest insults he could come up with – a fact that, in itself, scared him. Meanwhile, another flame appeared on his screen. This time it was worse. Before his eyes some of his data turned into gibberish. John Seabrook had a panic-attack . . . what about that last line of the first flame? *One good worm deserves another!* Had someone sent him a worm that was destroying all his information?

He backed up his files. Removed the disks from his computer. He sat in front of his screen, regarding it with suspicion and fear.

> I had the odd sensation that my computer was my brain, and my brain was ruined, and I was standing over it looking down at the wreckage. In my excitement over the new medium, I had not considered that in going on-line I was placing my work and my most private musings only inches from a roaring highway of data . . . and, like most highways, it didn't care about me. After thinking about this for a while, I noticed I was sitting in the dark, so I got up and pulled the chain on the floor lamp, and the bulb blew out. I thought, Wait a second, if my computer is connected to the outlet, is it possible that the worm could have gone into the plug and through the wall circuit and come out in the light bulb?
>
> The worm had entered my mind.[13]

While John Seabrook is undoubtedly "playing to his audience", the issue he raises warrants serious consideration. Increasingly, we are going to have to think about the relationship between human beings and thinking machines: about our ability to use a computer as an extension of our minds, memories and even our physiological abilities. (In her stunning novel, *Body of Glass* or *He, She or It*, Marge Piercy[14] blurs the boundaries when she portrays human beings with extensive computerised parts, and robots with software that has been programmed to make them the most understanding of human beings. So what's human and what's a machine – and what does it matter?)

For people like Mike Saenz, the creator of *Virtual Valerie*, there's no doubt about the way man will relate to machines in the cyberworld. "I think lust motivates technology," he says. "The first personal robots, let's face it, are not going to bring people drinks."[15] He goes on to give his version of cyber-society.

---

12. Seabrook, "My First Flame", p. 71.
13. Seabrook, "My First Flame", p. 73.
14. Marge Piercy, 1991, *He, She or It*, Knopf, New York, and *Body of Glass*, Penguin, London.
15. Mike Saenz, 1992, *Mondo 2000*, HarperCollins, New York, p. 275.

When I explain virtual reality to the uninitiated, they just don't get it. But they warm immediately to the idea of virtual sex . . . I have a silly idea for a product called Strip Teacher. She goes, "Tell me the name of the thirteenth president of the United States and I'll show you my tits."[16]

Cheris Kramarae, one of the new breed of data critics, sees this preoccupation with dehumanised sex in cyberspace as extremely worrying. It makes sex a computer sport – like hunting, and shooting tanks, and capturing aliens. Sex "will again be removed from intimacy as many women know it or wish it," she comments.[17]

She is not making a dismissable, prudish Victorian response to sex, but seeks to point out that virtual porn, sex robots, and pornware interactive games are a very different thing from the personal relationships that women have indicated as their preference.[18] Cheris Kramarae wants to make it clear that we cannot underestimate the extent to which sex is becoming the new medium's message.

"In France, the national fiber-optic Minitel system [which put a computer in every home] is largely funded by phone sex."[19] This pushes us to ask – is this the society, the community, the personal values that we want? Or should we have some different priorities for social policy in cyberspace?

Howard Rheingold, the mastermind of virtual reality, has his own view on this topic. To him, the future is less a matter of the reduction in intimacy, and more about disembodiment.

The secondary social effects of techno-sex are potentially revolutionary. If technology enables you to experience erotic frissons or deep physical, social, emotional communion with another person, with no possibility of pregnancy or sexually transmitted disease, what then of conventional morality, and what of the social rituals and codes that exist solely to enforce that morality? Is disembodiment the ultimate sexual revolution and/or the first step toward abandoning our bodies?[20]

To Howard Rheingold it is crystal clear; the culture is going through a revolutionary change. Everything we know and every value we take for granted is up for renegotiation; and this includes sex, humanity, and perception. "Privacy and identity and intimacy will become tightly coupled with something we don't have a name for yet," he says,[21] but it will combine the personal with the virtual. It will lead to new notions of entertainment and of ecstasy.

To Howard Rheingold, cyberspace and techno-sex will ensure that sexual satisfaction is available to one and all. Everyone will be able to have techno-orgasms whenever

---

16. Saenz, *Mondo 2000*, p. 274.
17. Kramarae, "A Backstage Critique of Virtual Reality", p. 47.
18. See Shere Hite, 1987, *Women and Love: A Cultural Revolution in Progress*, Knopf, New York.
19. Suzanne Stefanac, 1993, "Sex and the New Media", *New Media*, April, pp. 38–45.
20. Howard Rheingold, 1991, *Virtual Reality*, Secker & Warburg, London, pp. 351–2.
21. Rheingold, *Virtual Reality*, p. 352.

and as often as they desire. Which could be another version of paradise; but it could also be a means of keeping a community in line. As he says – somewhat cavalierly – "perhaps cyberspace is a better place to keep most of the population relatively happy, most of the time".[22]

According to Howard Rheingold, the most FAQ about virtual reality is whether it will become a form of electronic LSD, a technology with mind-altering capacities. While he acknowledges the brainwashing possibilities of the medium, they do not seem to him to be the major problem.

> What seems to me a much more credible fear [is] that VR might become as addictive, energy-sapping and intellect-dulling as *television*, "the plug-in drug" that requires the average American abuser to consume for more than seven hours a day, to the profit of those who won the battle for control of the gateway.[23]

This is, of course, another variation of the information-rich and poor: the former control the gateway and produce the technological "opiate", and the latter are led along the ecstasy path. Not that Howard Rheingold seems unduly perturbed by the scenario he describes. So what's wrong with electronic LSD, he asks? The new technologies simply mean that "people will use cyberspace to get out of their minds as well as their bodies".[24]

Think about it. And about the policies we want for cyber-society.

A Toast to History

No doubt the world looked all gloom and doom to the nuns and the monks who saw their marvellous manuscript culture being cast aside for the coarse sheets of crude print. As they saw their skills disparaged, their learning mocked, their truths challenged, they genuinely believed they were witnessing the corruption of the human soul and the end of civilised values. They could no more predict the marvels that would come with the flowering of print culture than some of us can predict the human enhancement that will occur when the full potential of net culture is realised.

I must confess that, despite my great enthusiasm for my computer and all that it allows me to do, I feel a distinct affinity with the nuns and the monks when I read some of the accounts of techno-sex and "condom emotionalism".[25] From where I stand in print culture, it looks like corruption of the psyche and the end of humanist values. And that's not my only point of identification with the scribes of the past.

As I work, I am still surrounded by books. There is not a spare inch of wall-space in the house. These books tell my life story: from the novels I studied at university (and

---

22. Rheingold, *Virtual Reality*, p. 352.
23. Rheingold, *Virtual Reality*, p. 355.
24. Rheingold, *Virtual Reality*, p. 356.
25. The term used by Allen Ulrich, 1992, to describe virtual sex: see "Here's Reality", *San Francisco Examiner*, 7 March.

the hundreds I read in London), to the enormous collection of women's studies publications that marked the success of the 1980s, right through to a good and recent collection of books on cyber-culture.

But I have to acknowledge that, these days, books are the background to my life. Apart from a very few, I don't consult them much any more. Gone is the time when I would have hundreds of them open, all over the place; when I couldn't go a day without my "shot" of Virginia Woolf, or Germaine Greer, or Aphra Behn. But now they stay on their shelves. I don't co-exist with them as I used to. They aren't my reference system; they aren't my gateway to another intellectual realm.

They aren't even specially insured as they once were; these days it is my computer for which I pay the premium.

Of course, I still buy books. By the hundreds. Nearly all novels. And there are stacks of good books next to my bed which I read at night. But only after I have been nattering on the net; and only for a comparatively short time.

This is why I am acutely conscious of my position as a member of the generation in transition. I am someone for whom books – and writing and reading – have been a way of life. Yet each day now, I know that books – and writing and reading – are becoming less and less central to my research, my work, and my leisure pursuits.

And this is a loss. It is a very big gap. It gives me some idea of how the nuns and the monks felt when the importance of their manuscripts continued to diminish and the book took their place as the primary information medium.

Some people now are beginning to feel the same way about paper and the written record.

Historians, for example, are having to face the possible end of written history. The paper documents that have been their source of information are giving way to electronic archives. There is nothing to touch or smell; there is no tangible sign of where the records come from or whose they were. The discussion at a seminar, History in the Age of the Computer, (Van Leer Institute, Jerusalem, 1995) was pessimistic.

> "It is time to consider the possibility that we are facing the demise of history writing as an academic profession," said Prof. Anita Shapira of Tel Aviv University . . . On the one hand, she noted, the historian is confronted by an overwhelming outpouring of electronically delivered information. On the other, some of the most revealing sources are disappearing from social convention.
>
> "Letters are rarely written," she noted. "Diaries are almost extinct."[26]

In many ways, electronic archives are no substitute. They don't necessarily inspire the same commitment as diaries, letters, and printed documents have done.

> Archivists would protect with their lives a historic parchment handed down though the centuries. But would a computer technician feel the same way about a data tape or CD-ROM?

---

26. Abraham Rabinovich, 1995, "Historians Told to Byte the Papyrus", *Age*, 30 May, p. 29.

"Now there is no sentiment," said [Dr Ronald Zweig, who is at the cutting edge of the new technologies]. "After five years, the electronic archivists throw it out. They don't have the culture of archive keeping. It's not taught in computer courses."[27]

Yet even as I note the losses, I sense the gains. I know what it is like to have access to the net; I know how school and university students are responding to the thrill of doing their own information. I look forward to the electronic equivalent of the novel, and I enjoy the services of the Internet café.

"Netcafé" (St Kilda, Melbourne) serves great cappuccinos and deli food, while upstairs there are six PCs "with graphic capability, running the Netscape World Wide Web browser, plus four text-only terminals".

Netcafé's proprietor, Mr Gavin Murray, said, "One of the disks which will be updated each night is devoted to storing a large quantity of popularly accessed World Wide Web pages to alleviate time delays associated with downloading large graphic files." . . .

"We are expecting all types of patrons, from small business users who may want to use the e-mail facilities, to the real hacker types," he said.

"This will be a meeting place for all those interested in the net to exchange ideas at all hours of the day and night."[28]

That the print-culture world is going is abundantly clear. With it go many of the conventions that we have cherished about who we are, what we know, and how we make sense of our universe. Cyber-society will be as dramatically different to us as print culture was to those of the manuscript era. New truths are in the making.

In the words of Amy Bruckman, the one woman thinker who along with six men was chosen to discuss the significance of the new technologies for *Technology Review*:

One of the really exciting things about what is going on is that we're moving away from the idea that truth is contained only in libraries and official databases. People are realizing that truth is created by communities of people. Here's a personal example. I keep tropical fish, and sometimes I need some information about their care and feeding. Yes, I could go to the Library and pore through a book on tropical fish – but I'll probably never find the right answer to my question. Or I can post a message on the Usenet group, alt.aquaria, and someone will respond in three minutes with exactly the information I need. The network is changing our basic notion of the nature of information. We can't think of information and community as separate concepts any more.[29]

That's the challenge for cyber-policy of the future.

---

27. Rabinovich, "Historians Told to Byte the Papyrus", p. 29.
28. Stan Beer, 1995, "Internet Cafes Cater to Wide Tastes", *The Australian*, 11 April, p. 42.
29. Amy Bruckman, 1994, "Seven Thinkers in Search of an Information Highway", *Technology Review*, August–September, p. 45, my emphasis.

Bibliography

Abbott, Joell, 1995, "Computer Phobic Tale", *geekgirl*, 1, p. 3.

Adams, Geoffrey, 1993, "The Shape of Things to Come", *ALCS News*, 6, July, p. 2.

Allan, Bonnie, 1994, "Right Beat, Wrong Angle", *MS Magazine*, January–February, pp. 92–3.

Alloway, Nola, 1995, *The Construction of Gender in Early Childhood*, Curriculum Corporation, Carlton, Vic.

Atkinson, Peter, 1995, "Court Told of On-line Lust", *Sunday Mail*, 12 February, p. 46.

Atwood, Margaret, 1994 "Not Just a Pretty Face", *Women's Review of Books*, 10, 4, January, pp. 6–7.

Aubert, Eric, 1993, *Campus Review*, 15–21 April, p. 1.

Barry, John A., 1993, *Technobabble*, MIT, Cambridge, Mass.

Beer, Stan, 1995, "Internet Cafés Cater to Wide Tastes", *The Australian*, 11 April, p. 42.

Belenky, Mary Field, *et al.*, 1986, *Women's Ways of Knowing*, Basic Books, New York.

Benyon, J., 1990, "A School for Men: Learning Macho Values at School", paper given at conference of British Association for the Advancement of Science, Swansea University, 24 August.

Bernard, Jessie, 1973, "My Four Revolutions: An Autobiographical History of the ASA", in Joan Huber (ed.), *Changing Women in a Changing Society*, University of Chicago Press.

Billington, James H., 1990, *The Electronic Erosion of Democracy*, Mortenson Distinguished Lecture, University of Illinois Library, Urbana-Champaign.

Bodine, Ann, 1975, "Androcentrism in Prescriptive Grammar", *Language in Society*, 4, 2, pp. 129–56.

Bolter, Jay David, 1993, "Alone and Together in the Electronic Bazaar", *Computers and Composition*, 10, 2, p. 11.

Brand, Stewart (1987), 1988, *The Media Lab: Inventing the Future at MIT*, Penguin, New York.

Bright, Martin, 1995, "On the Internet: the First Holy War", *Guardian*, reprinted *Sydney Morning Herald*, 11 March, p. 23.

Brown, Paul, 1993, "Digital Technology and Motion Pictures", in Ross Harley (ed.), *New Media Technologies*, Australian Film, Television and Radio School, Sydney.

Bruck, Jan, 1991, "Writing in the Electronic Age", *Media Information Australia*, 61, p. 51.

Bruckman, Amy, 1993 "A Study in MUDs", *Wired*, July–August, pp. 70–1.

——, 1994, "Seven Thinkers in Search of an Information Highway", *Technology Review*, August–September, p. 45.

Bulkeley, William M., 1994, "A Tool for Women, a Plaything for Men", *Wall Street Journal*, reprinted *Age*, 5 April.

Burgess, David, 1993, "Hardcore Software", *The Australian*, 1 June, p. 9.

Butterworth, Dianne, 1993, "Wanking in Cyberspace", *Trouble and Strife*, 27, Winter, pp. 33–7.

Capek, Mary Ellen, 1987, *A Woman's Thesaurus: An Index of Language Used to Describe and Locate Information by and about Women*, Harper and Row, New York.

Carli, Donald, 1994, "A Designer's Guide to the Internet", *Step-by-Step Graphics* (Peoria, Ill.), November–December, p. 27.

Chester, Rodney, 1995, "A Place in Cyberspace", *Courier Mail*, 13 April, p. 22.

Cohen, Roger, 1991, "Weak Sales Put a Damper on an Annual Book Party", *New York Times*, 27 May, pp. 119, 224.

Cole, Anne, Tom Conlon, Sylvia Jackson & Dorothy Welch, 1994, "Information Technology and Gender: Problems and Proposals", *Gender and Education*, 6, 1, p. 78.

Computers and Composition; A Journal for the Teachers of Writing, University of Illinois and Michigan Technological University.

Cotton, Bob, & Richard Oliver, 1993, *Understanding Hypermedia*, Phaidon Press, London.

Cox, David, 1994, "Women Tearing down the Walls of the Male Games Bastion", *The Age*, 6 December, p. 29.

Crawford, Frank, 1995, "Pornshops on the Internet Not Easily Closed Down", *The Australian*, 28 March, p. 39.

Cross, Rosie, 1995, "Cyberfeminism: Interview with Sadie Plant", *geekgirl*, 1, p. 4.

——, 1995, "Modem Grrrl: Wicked St Jude – Interview with Rose X", *geekgirl*, 1, p. 8.

Daly, Mary, 1973, *Beyond God the Father*, Beacon Press, Boston.

Darnton, Robert, 1986, "First Steps Towards a History of Reading", *Australian Journal of French Studies*, 22, p. 5.

Davies, Lynda, 1994, "The Gendered Language of Technology", paper given at conference of International Communication Association, Sydney, 11–15 July.

Davis, Eric, 1992, "Cyberlibraries", *Lingua Franca*, February–March.

de Beauvoir, Simone, 1972, *The Second Sex*, Penguin, London.

Delaney, Paul, & George Landow (eds), 1991, *Hypermedia and Literary Studies*, MIT Press, Cambridge, Mass.

Dettman, Pam, 1993, "A Laptop Revolution", in Irene Grasso & Margaret Fallshaw (eds), *Reflections of a Learning Community*, Methodist Ladies' College, Melbourne.

Dibbell, Julian, 1994, "Data Rape: A Tale of Torture and Terrorism On-line", in *Sydney Morning Herald Good Weekend*, 19 February, p. 34.

Dorner, Jane, 1993, "Whose Text Is It Anyway?", *Times Higher Education Supplement*, 23 April.

Dunleavy, Maurice, 1993, "The Death of the Book", *Blast*, 21, Autumn, pp. 10–11.

Easlea, Brian, 1980, *Witch-hunting, Magic and the New Philosophy: An Introduction to the Debates of the Scientific Revolution, 1450–1750*, Harvester Press, Sussex.

Ebben, Maureen, 1992, "In Bed with Electronic . . .", Internet communication, 22 November.

——, & Cheris Kramarae, 1993, "Women and Information Technologies: Creating a Cyberspace of Our Own", in H. Jeanie Taylor, Cheris Kramarae & Maureen Ebben (eds), *Women, Information Technology and Scholarship*, Center for Advanced Studies, University of Illinois, Urbana-Champaign.

Eisenstein, Elizabeth, 1983, *The Printing Revolution in Early Modern Europe*, Cambridge University Press, Cambridge.

Fallshaw, Margaret, 1993, "The Promises of Educational Technology", in Irene Grasso & Margaret Fallshaw (eds), *Reflections of a Learning Community*, Methodist Ladies' College, Melbourne.

Farmer, Monique, 1993, "PCs Can Be Cool", *Sydney Morning Herald*, 8 February.

Febvre, Lucien, & Henri-Jean Martin, 1984, *The Coming of the Book*, Verso, London.

Forester, Tom (ed.), 1989, *Computers in the Human Context*, Basil Blackwell, Oxford.

Fraser, Antonia, 1984, *The Weaker Vessel: Women's Lot in Seventeenth-century England*, Weidenfeld & Nicolson, London.

Gage, Matilda Joslyn (1893), 1980, *Women, Church and State: The Original Exposé of Male Collaboration against the Female Sex*, Persephone Press, Watertown, Mass.

Gambon, Jill, 1995, "Networks Wary of Anti-porn Bill", *The Australian*, 28 February.

Garton, Andrew, 1994, "*The Net – Promise or Threat?*", Internet communication, 19 February, <agarton@peg.apc.org>.

George, Gwen, & Linda Adamson, 1995, *Information Superhighway or User-friendly Byway? The Telephone and Older Women*, Older Women's Network, Sydney.

Gilbert, Sandra M., & Susan Gubar, 1980, *The Madwoman in the Attic; The Woman Writer in the Nineteenth-century Literary Imagination*, Yale University Press.

Goldberg, Philip, 1974, "Are Women Prejudiced against Women?", in Judith Stacey *et al.* (eds), *And Jill Came Tumbling After: Sexism in American Education*, Dell Publishing, New York.

Graham, Alma, 1975, "The Making of a Non-sexist Dictionary", in Barrie Thorne & Nancy Henley (eds), *Language and Sex: Difference and Dominance*, Newbury House, Mass.

Grasso, Irene, & Margaret Fallshaw (eds), 1993, *Reflections of A Learning Community: Views on the Introduction of Laptops at MLC*, Methodist Ladies' College, Melbourne.

Green, Eileen, Jenny Owen & Den Pain (eds), 1993, *Gendered by Design? Information Technology and Office Systems*, Taylor & Francis, London.

Gregorian, Vartan, 1992, "Technology, Scholarship and Humanities: The Impact of Electronic Information", keynote address given at conference of National Academy of Science and Engineering, Irvine, Calif., 30 September–2 October.

Guldberg, Hans Hoegh, 1990, *Books – Who Reads Them?*, Australia Council, Sydney.

Gunn, Angela, 1991, "Computer Bulletin Boards Not Just Boy Toy", *New Directions for Women*, November–December, p. 7.

Hafner, Katie, & John Markoff, 1993, *Cyberpunk: Outlaws and Hackers on the Computer Frontier*, Corgi Books, London.

Haraway, Donna, 1985, "Manifesto for Cyborgs", *Socialist Review*, 80, pp. 65–108.

——, 1987, "Animal Sociology and a Natural Economy of the Body Politic", in Sandra Harding & Jean O'Barr (eds), *Sex and Scientific Inquiry*, University of Chicago Press.

Harley, Ross, 1993, "The Nature of New Media Technologies", in Ross Harley (ed.), *New Media Technologies*, Taking Care of Business Series, Australian Film, Television and Radio School, Sydney.

—— (ed.), 1993, *New Media Technologies*, Taking Care of Business Series, Australian Film, Television and Radio School, Sydney.

Harrington, Paul (in Hong Kong), 1995, "Cybersurfers Enjoy Anarchy", *The Australian*, 11 April, p. 37.

Haynes, Colin, 1993, *The Electronic Author*, 2, Winter, p. 2.

Herring, Susan, Deborah Johnson, & Tamra DiBenedetto, 1992, "Participation in Electronic Discourse in a Feminist Field", in Kira Hall, Mary Bucholz & Birch Moonwomon (eds), *Proceedings of the Second Berkeley Women and Language Conference*, University of California, Berkeley.

Heyward, Michael, 1993, *The Ern Malley Affair*, University of Queensland Press, St Lucia, Qld.

Hiley, Michael, 1993, "Writing for Multimedia", *The Electronic Author*, 2.

Hill, Stephen, 1988, *The Tragedy of Technology*, Pluto Press, London.

Hiscock, John, 1993, "The Chips Get down to Sex", *Herald-Sun*, 2 October, p. 22.

Hite, Shere, 1987, *Women and Love: A Cultural Revolution in Progress*, Knopf, New York.

Holderness, Mike, 1993, "Down and Out in the Global Village", *New Scientist*, May, pp. 36–40.

——, 1994, "On-line Harassment Slips through the Net", *The Australian*, 19 January, p. 18.

——, 1994, "Netiquette for the Novices", *Guardian*, reprinted *Age Green Guide*, 13 October, p. 25.

House of Representatives Standing Committee on Long Term Strategies, 1991, *Australia as an Information Society: Grasping New Paradigms*, Parliament of the Commonwealth of Australia.

Howell, William, 1994, "Point of View", *Chronicle of Higher Education*, 8 June, p. A40.

Hughes, Bob, 1993, "The Parody Machine", *The Electronic Author*, 2, p. 7.

Hughes, Robert, 1993, "Introduction", in Michael Heyward, *The Ern Malley Affair*, University of Queensland Press, St Lucia, Qld.

Johnstone, Craig, 1995, "Surfin' for Sex", *Courier Mail*, 5 April, p. 9.

Jones, J., 1995, letter, *The Australian*, 11 April, p. 26.

Jones, Randye, 1993, "Netters' Behavior and Flame Wars", On-line communication, September.

Jones, Steven G. (ed.), 1994, *Cybersociety: Computer-mediated Communication and Community*, Sage, Newbury Park, Calif.

Kantrowitz, Barbara, 1994, "Men, Women and Computers", *Newsweek*, 16 May, p. 51.

Kaplan, James, 1989, "Inside the Club", *New York Times Magazine*, 11 June, p. 62.

Kaplan, Nancy, & Eva Farrell, 1994, "Weavers of Webs: A Portrait of Young Women on the Net", *Arachnet Electronic Journal on Virtual Culture*, 2, 3, July.

Kelly, Kevin, & Howard Rheingold, 1993, "The Dragon Ate My Homework", *Wired*, July–August, p. 73.

Kernan, Alvin, 1990, *The Death of Literature*, Yale University Press.

Kiesler, Sara, Lee Sproull, & Jacqueline S. Eccles, 1985, "Poolhalls, Chips and Wargames: Women in the Culture of Computing", *Psychology of Women Quarterly*, 1, 4, pp. 451–62.

Killick, David, & Sonya Sandham, 1993, "Policing the Techno-nasties", *Sydney Morning Herald Spectrum*, 18 December, p. 1A.

Kingson, Jennifer, 1989, "Where Information Is All, Pleas Arise for Less of It", *New York Times*, 9 July, p. E9.

Kirkup, Gill, & Laurie Smith Keller (eds), 1992, *Inventing Women: Science, Technology and Gender*, Polity Press and Open University, Cambridge.

Kramarae, Cheris, 1994, "A Backstage Critique of Virtual Reality", in Steven G. Jones (ed.), *Cybersociety*, Sage, Newbury Park, Calif.

—— (ed.), 1988, *Technology and Women's Voices*, Routledge & Kegan Paul, New York.

——, & H. Jeanie Taylor, 1993, "Women and Men on Electronic Networks: A Conversation or a Monologue?", in H. Jeanie Taylor, Cheris Kramarae & Maureen Ebben (eds), *Women, Information Technology, and Scholarship*, Center for Advanced Studies, University of Illinois, Urbana-Champaign.

——, & Jana Kramer, 1995, "Net Gains, Net Losses", *Women's Review of Books*, February, pp. 33–5.

——, et al., 1985, *A Feminist Dictionary*, Pandora, London.

Kranzberg, Melvin, 1989, "The Information Age", in Tom Forester (ed.), *Computers in the Human Context*, Basil Blackwell, Oxford.

La France, Marianne, & Anne Meyer, 1992, "Computers as Barrier or Vehicle to Equity?", paper presented at National Conference on Computing and Values, New Haven, Conn., August, <LAFRANCE@BCVMS.bitnet>.

Landow, George P., & Paul Delany, 1991, "Hypertext, Hypermedia and Literary Studies: The State of the Art", in George Landow & Paul Delany (eds), *Hypermedia and Literary Studies*, MIT Press, Cambridge, Mass.

Landwehr, Larry, 1993, "Some Comments on the *London Times Educational Supplement* article", *Computer Underground Digest*, April.

Leebaert, Derek (ed.), 1991, *Technology 2001: The Future of Computing and Communications*, MIT Press, Cambridge, Mass., and London.

Levy, Steven, 1994, *Hacker: Heroes of the Computer Revolution*, Penguin, London.

Lines, Michael, 1995, letter, *The Australian*, 21 March, p. 24.

Loader, David, 1993, "Reconstructing an Australian School", in Irene Grasso & Margaret Fallshaw (eds), *Reflections of a Learning Community*, Methodist Ladies' College, Melbourne.

Lyman, Peter, 1994, Follett Lecture, Internet, 4 October, directed courtesy of Janine Schmidt and Colin Steele.

Lyon, David, 1994, *The Electronic Eye: The Rise of Surveillance Society*, University of Minnesota Press, Minneapolis.

Mackay, Hugh, 1992, "Watched Any Good Books Lately?", *Bulletin*, 11 August, p. 45.

Marcure, Judy, 1993, "Dynamic Documents and Print Conventions: New Directions in Scholarly Communications", private communication.

Margie, 1993, "Hi Girls, See You in Cyberspace", *Sassy*, May, pp. 72–3, 80.

Martyna, Wendy, 1983, "Beyond the He/Man Approach", in Barrie Thorne, Cheris Kramarae & Nancy Henley (eds), *Language, Gender and Society*, Rowley, Mass., pp. 25–37.

McCaughan, David, & Greg Wrobel, 1993, *Out of the Real: Teens and Technology*, McCann-Erickson, Sydney.

McIntosh, Trudi, 1993, "How to Take the Electronic Leap . . .", *Australian Weekend Review*, 10–11 April, p. 4.

——, 1994, "Copyright Chaos in 'Data Soup'", *The Australian*, 12 April, p. 24.

——, 1995, "Cyberspace Heroine Shoots for the Stars", *The Australian*, 28 March, p. 23.

McLaughlin, Margaret L., Kerry C. Osborne & Christine B. Smith, 1995, "Standards of Conduct on Usenet", in Steven G. Jones (ed.), *Cybersociety*, Sage, Newbury Park, Calif.

McLuhan, Marshall, 1974, *Understanding Media*, Abacus, London.

Mead, Margaret, 1950, *Male and Female*, Gollancz, London; reprinted 1971, Penguin, Harmondsworth.

Melody, Bill (ed.), 1992, "The Intelligent Telecommunication Network: Privacy and Policy Implications of Calling Line Identification and Emerging Information Services", *Proceedings of a CIRCIT Conference*, Centre for International Research on Communication and Information Technologies, Melbourne.

Miller, Casey, & Kate Swift, 1976, *Words and Women*, Anchor/Doubleday, New York.

Mitchell, Lisa, 1995, "Princess Internet Takes her Glass Wings to the Ball", *Age*, 7 March, p. 27.

Moran, Albert (ed.), 1990, "The Media of Publishing", *Continuum: An Australian Journal of the Media*, 4, 1.

Morris, Judith, 1989, *Women in Computing*, Computer Weekly Publications, Surrey.

Moyal, Ann, 1995, "Crisis in Communication Research", paper presented at the Communication Research Centre, *The Informationless Society Seminar*, 16 February, p. 6.

Mulvaney, John, & Colin Steele (eds), *Changes in Scholarly Communication Patterns: Australia and the Electronic Library*, Australian Academy of the Humanities, Canberra, Occasional Paper no. 15.

Negroponte, Nicholas, 1995, *Being Digital*, Knopf, New York.

Painter, Joanne, 1993, "Uni to Look into Sexual Harassment via Computer", *Age*, 5 November, p. 8.

Papert, Seymour, 1993, *The Children's Machine: Rethinking in the Age of the Computer*, Basic Books, New York.

Pendergast, Mark, 1989, "Sandy Berman: A Man for All Subjects", *Wilson Library Bulletin*, March, p. 50.

Penelope, Julia, 1990, *Speaking Freely*, Athene, New York.

Pereira, Joseph, 1994, "Video Games: The Chips Are Stacked in Favor of Boys", *Wall Street Journal*, reprinted *Age*, 5 April, p. 21.

Piercy, Marge, 1992, *He, She, and It* (USA) and *Body of Glass* (UK and Australia), Penguin.

Pitkow, James, & Mimi Recker, Georgia Institute of Technology: see Donald Carli, 1994, "A Designer's Guide to the Internet", *Step-by-Step Graphics* (Peoria, Ill.), November–December, p. 27.

Postman, Neil, 1986, *Amusing Ourselves to Death: Public Discourse in the Age of Show Business*, Penguin, New York.

Powell, Gareth, 1994, "The Path to Compact Learning", *Sydney Morning Herald*, 28 March, p. 45.

Price, Jenna, 1994, "When One Reader's Crit is Another's Bull", *Sydney Morning Herald*, 8 January, p. 2A.

Pringle, Peter, 1993, "Video Games Link Bloody Fantasy to Violent Reality", *Age*, 22 September, p. 15.

Rabinovich, Abraham, 1995, "Historians Told to Byte the Papyrus", *Age*, 30 May, p. 29.

Rakow, Lana F., 1992, *Gender on the Line: Women, the Telephone and Community Life*, University of Illinois Press, Urbana and Chicago.

Reineke, Ian, 1993, "Fear, Greed and the Electronic Library", in John Mulvaney & Colin Steele (eds), *Changes in Scholarly Communication Patterns*, Australian Academy of the Humanities, Occasional Paper no. 15.

Rheingold, Howard, 1986, "Virtual Communities", *Whole Earth Review*, Summer.

——, 1991, *Virtual Reality*, Secker & Warburg, London.

——, 1993, "A Slice of Life in My Virtual Community", in L. M. Harasim (ed.), *Global Networks: Computers and International Communication*, MIT Press, Cambridge, Mass.

Rich, Adrienne, 1979, *On Lies, Secrets and Silence*, Virago, London.

Ricketson, Matthew, 1994, "Reading the Market", *Flight Deck Magazine*, March, pp. 12–13.

Robinson, Carol, 1993, "Publishing's Electronic Future", *Publishers' Weekly*, 6 September, p. 46.

Rossman, Parker, 1992, *The Emerging Worldwide Electronic University: Information Age Global Higher Education*, Greenwood Press, Westport, Conn.

Roszak, Theodore, 1994, *The Cult of Information*, University of California Press, Berkeley.

Rushkoff, Douglas, 1994, *Cyberia: Life in the Trenches of Hyperspace*, Harper, San Francisco.

Saenz, Mike, 1992, *Mondo 2000*, HarperCollins, New York.

Sanders, Jo Shuchat, & Antonia Stone, for the Women's Action Alliance, 1986, *The Neuter Computer: Computers for Girls and Boys*, Neal Schuman Publishers, New York.

Schiebinger, Londa, 1991, *The Mind has No Sex?*, Harvard University Press, Cambridge, Mass.

Seabrook, John, 1994, "My First Flame", *New Yorker*, 6 June.

Seaman, Bill, 1993, "Interactive Videodisk Production", in Ross Harley (ed.), *New Media Technologies*, Australian Film, Television and Radio School, Sydney.

Sharples, Mike, 1993, "Beyond the Word Processor", *The Electronic Author*, 1, p. 12.

Shneiderman, Ben, 1989, "Designing the User Interface", in Tom Forester (ed.), *Computers in the Human Context*, Basil Blackwell, Oxford.

Shoemaker, Pamela J., 1991, *Gatekeeping*, Sage, Newbury Park, Calif.

Simper, Errol, 1995, "A Matter of Time as Pay TV Faces Inbuilt Problems of the Market", *The Australian*, 18 January, p. 2.

Spender, Dale, 1980 (1985), *Man Made Language*, Routledge, London.

——, 1982, *Women of Ideas – And What Men Have Done to Them*, Pandora, London.

——, 1986, *Mothers of the Novel: 100 Good Women Writers Before Jane Austen*, Pandora, London.

——, 1989, *The Writing or the Sex?*, Athene, New York.

——, 1990, *Invisible Women: The Schooling Scandal*, The Women's Press, London.

—— (ed.), 1991, *Heroines*, Penguin, Ringwood, Vic.

—— (ed.), 1992, *Living by the Pen*, Athene, New York.

——, & Lynne Spender (eds), 1983, "Gatekeeping: The Denial, Dismissal and Distortion of Women", *Women's Studies International Forum* (special issue), 6, 5.

Spender, Lynne, 1993, *Newswrite: Newsletter of the NSW Writers' Centre*, March–April, p. 4.

Stager, Gary, 1993, "Computers for Kids – Not Schools", in Irene Grasso and Margaret Fallshaw (eds), *Reflections of a Learning Community*, Methodist Ladies' College, Melbourne.

Stange, Eric, 1987, "Millions of Books are Turning to Dust – Can They Be Saved?", *New York Times Book Review*, 29 March, p. 3.

Stanley, Autumn, 1981, "Daughters of Isis, Daughters of Demeter", *Women's Studies International Quarterly*, 4, 3.

——, 1992, "Do Mothers Invent? The Feminist Debate on the History of Technology", in Cheris Kramarae & Dale Spender (eds), *The Knowledge Explosion*, Teachers College Press, New York.

Stansberry, Domenic, 1993, "Taking the Plunge", *New Media*, February, p. 35.

Starr, Victoria, 1994, *k.d. lang: All You Get is Me*, HarperCollins, London.

Stefanac, Suzanne, 1993, "Sex and the New Media", *New Media*, April, pp. 38–45.

Sterling, Bruce, 1992, *The Hacker Crackdown: Law and Order on the Electronic Frontier*, Penguin, London.

——, 1995, "Good Cop, Bad Hacker", *Wired*, May, p. 122.

Stoll, Clifford, 1995, *Silicon Snake Oil: Second Thoughts on the Information Superhighway*, Bantam Doubleday, New York, p. 3.

Taylor, H. Jeanie, Cheris Kramarae & Maureen Ebben (eds), 1993, *Women, Information, Technology and Scholarship*, Center for Advanced Study, University of Illinois, Urbana-Champaign.

Taylor, Mark C., & Esa Saarinen, 1994, *Imagologies: Media Philosophy*, Routledge, London.

Tinkler, Don, Tony Smith, Peter Ellyard & David Cohen, 1994, *Effectiveness and Potential State-of-the-Art Technologies in the Delivery of Higher Education*, Commonwealth of Australia.

Todd, Janet, 1984, *A Dictionary of British and American Women Writers, 1660–1800*, Methuen, London.

Tucker, Debbie, 1995, "Spider Spins Web for Women", *Courier Mail*, 13 April, p. 22.

Turek, Kathleen A., & Judith Hudson, 1994, "The Future is Here", *Feminist Collections*, 5, 3, Spring, p. 13.

Turkle, Sherry, 1988, "Computational Reticence: Why Women Fear the Intimate Machine", in Cheris Kramarae (ed.), *Technology and Women's Voices*, Routledge & Kegan Paul, London.

Verrender, Ian, 1994, "The Thought Police", *Sydney Morning Herald Spectrum*, 26 March, p. 4A.

Waldren, Murray, 1994, "In the Blood", *The Australian Magazine*, March, p. 19.

Wales, Elspeth, 1995, "MPs Attempt to 'Feminise' IT as Participation Rate Declines", *The Australian*, 4 April, p. 27.

Walkerdine, Valerie, 1988, *The Mastery of Reason*, Routledge, London.

Wark, McKenzie, 1994, "The Video Game as an Emergent Media Form", *Media Information Australia*, 71, February, pp. 21–30.

Whitney-Smith, Elin, 1991, "Information Doesn't Want", *Whole Earth Review*, Fall, pp. 38–40.

Willard, Charity, 1984, *Christine de Pizan*, Persea Books, New York.

Williams, Frederick, & John V. Pavlik, 1994, *The People's Right to Know: Media, Democracy and the Information Highway*, Lawrence Erlbaum, New Jersey.

Williams, Raymond, 1978, *The Long Revolution*, Penguin, London.

Willis, Sue, 1991, "Girls and their Futures", keynote address given at National Conference of Australian Women's Education Coalition, Perth, 7–9 June.

Wilson, David, 1993, "Information Technology: A Computer Program that Can Censor Electronic Messages Sets off a Furor", *Chronicle of Higher Education*, 12 May, p. A21.

Woodmansee, Martha, 1984, "The Genius of Copyright: Economic and Legal Conditions of the Emergence of the Author", *Eighteenth Century Studies*, 4, p. 425.

Wylie, Margie, 1995, "No Place for Women", *Digital Media*, 4, 8, January.

Young, Luke T., *et al.*, 1988, "Academic Computing in the Year 2000", *Academic Computing*, 2, 7, May–June, pp. 8–12, 62–5.

Index